Goethe on Science

Goethe on Science

A selection of Goethe's writings
edited and introduced by
Jeremy Naydler

Floris Books

First published in English in 1996 by Floris Books
Second impression 2000

British Library CIP Data available

ISBN 0-86315-237-6

Printed in Great Britain
by Cromwell Press, Trowbridge

Contents

When we venture into knowledge and science, we do so only to return better equipped for living.

Johann Wolfgang von Goethe

Acknowledgments

The idea of creating a compilation of Goethe's writings on scientific method arose through working in small groups with Goethe's methodology over a number of years. In many respects this book is the fruit of that group-work. I hope that it may also act as the seed for similar groups in the future. I am especially indebted to Michael Preston, Christina Creek, Stewart Richards, Joy Mansfield and Robert Chamberlain for their sustained commitment to exploring the Goethean approach to nature, and for providing the original stimulus for this book; to Sam Betts for many profound discussions, and for his helpful comments on early drafts of the manuscript; and to Claudia Böse for her patient encouragement and invaluable help with the translations. Thanks are also due to Vicky Yakehpar and Sue Wasley for their willing assistance with the preparation of the manuscript; and finally, to Iain and Suzie Geddes, Felix Padel, Lovanne Richards and Ajit Lalvani for their recognition of the importance of Goethe, their friendship and support of this project.

This book is dedicated to all who love nature and are seeking to deepen their awareness of the natural world.

JN

Preface

This seems to be a time of reappraisal for Goethe's science. Several books have appeared in recent years which reopen the question of the validity of Goethe's approach to scientific investigation as this is exemplified in his work on colour, the metamorphosis of living form, and, to a lesser extent, geology and meteorology. Previously, the majority opinion seems to have been that, with regard to his work on colour, he was a muddle-headed dilettante who made a fool of himself in opposing Newton. But thorough investigation of his work has shown that, far from being a dilettante, Goethe was a careful and accomplished observer of the phenomena. He understood the role of experiment and theory, and was particularly aware of the dangers which could follow from taking an uncritical attitude towards the latter. His writings on methodology here seem very relevant and modern. It has even been suggested that his understanding of the role of experiment in physical investigations was, in several ways, superior to Newton's understanding of the role that experiment played. The difference is that Goethe worked in the domain of quality, whereas Newton worked in that of quantity, and for reasons which have no fundamental justification the latter has been declared the true ground of nature. Quality has been relegated to a secondary place, dependent on quantity and without real being. Historically, this view arose from the identification of (quantitative) mathematics as the key to nature, together with the incorporation of the ancient philosophy of atomism into science. This was the historical deed of Galileo, Descartes, Newton, and a number of others, which has shaped the very form which our scientific understanding takes today.

Goethe researched thoroughly the history of the science of colour, and came to recognize for himself that science is intrinsically historical. He said 'We might venture the statement that the history of science is science itself.' He distinguished different 'ways of conceiving' *(Vorstellungsarten)*, and came to understand the role of such ways of conceiving in what is naively taken to be purely factual. Furthermore, he realized that the way of conceiving could not be separated from the content, that is, from *what* is seen. The way of conceiving is integral to scientific cognition. It is here that Goethe seems to leap forward a century and a half to the work of Thomas Kuhn and others in the early 1960's. As is well known, the new philosophy of science has radically altered our under-standing of what science is, and it is remarkable how well Goethe's understanding of science corresponds with this. From our vantage point today, we can see that it was those in the last century who said that Goethe had an inadequate conception of science, who themselves had the inadequate conception and not Goethe. Fur-thermore, in view of the discovery that modern science is a culturally based activity, incorporating factors which do not have any intrinsic foundation within some 'pure' science, then we have no reason to mistake decisions made in the seventeenth century for the absolute and final truth about nature. The discovery of the irreducible historical dimension of science as a cultural enterprise, can liberate us from the dogmatism of science. It can also reopen us to possibilities other that those which were chosen historically. It is at this point that we can encounter Goethe's science of colour as a genuine possibility. No longer in thrall to the idea that quantity alone is real, we are liberated to recognize the reality of quality *in its own right,* and hence to discover the possibility of a new kind of macroscopic science. This is the direction in which Goethe takes us: towards a science of the lived experience of phenomena, instead of mathematical abstractions. and microscopic explanations.

Goethe's biological work had a much better reception than his work on colour. He has been recognized as the author of the term 'morphology' to describe the investigation of organic form, and his influence on the school of transcendental anatomy in the nineteenth century has been fully accredited. But acceptance has its own problems, and all too often the biological viewpoint which is attributed to Goethe is not the Goethean one at all. He is taken to be a Platonist — no doubt the use of the term 'archetype' encourages this — whereas he was nothing of the kind. Goethe did not subscribe to the dualism of the two-world theory, with its notion of the archetype as 'one over many' which is separate from the multiplicity of particular phenomena. The whole of his scientific work was in the opposite direction to this kind of metaphysical dualism. Goethe's science was a non-metaphysical science — unlike mathematical physics, which conceives of a hidden reality *behind* appearances to explain them. Interpretations of Goethe's archetype which see it as an Ideal Plan for organisms, wrongly project his science into the very metaphysical tradition from which it, in fact, escapes. For Goethe, all is within the phenomenon and nothing behind. But this is far from being a naive empiricism, which sees the surface appearance as all there is to the phenomenon. To begin with the phenomenon is only partly visible, and the task of science is to make the inner-dimensionality of the phenomenon visible, so that it becomes wholly visible. Goethe's way of science is much closer, philosophically, to phenomenological seeing than it is to traditional empiricism.

Just as mistaken, in its own way, as the notion of the archetype as an Ideal Plan, is the other common misinterpretation of the archetype as a common ancestor. This view seems to be a consequence of seeing Goethe retrospectively through Darwinian spectacles. But the archetype is not to be identified with a primitive organism in time in this way. The many organisms do not develop from the archetype by the kind of mechanical causality according to which Darwin envisaged the evolution of organisms from a primitive ancestor. Darwin himself may have encouraged this misinterpretation of Goethe, by including a reference to him in a footnote added in a later edition of his *Origin of Species.* The archetype is for Goethe neither an abstract one behind the many, nor is it a single beginning in time. It requires a new way of seeing. It is a feature of this new way of seeing that it sees unity in an inside-out way to the abstract unity which is no more than what things have in common. This is only an external unity. The archetype is the inner dimension of the phenomenon, but if we begin to see in the mode of the archetype then the many appear within the one — not, of course, as parts of the one (which would be self-contradictory). Each is seen as being a manifestation of the one, and so this is a *concrete* mode of unity — multiplicity within unity, instead of unity abstracted from multiplicity. This concrete mode of unity is one and many at the same time. It permits diversity within unity, whereas the abstract mode of unity excludes diversity and allows only uniformity. The richness of Goethe's concrete way of seeing multiplicity in the light of unity is at the opposite pole to the poverty of the external way of seeing unity as sameness — which is what the notion of an Ideal Plan really comes down to in the end. But, to understand this, we have to experience Goethe's way of seeing difference in the light of unity for ourselves.

There is today a growing concern over the impact which science and technology are having on nature. Faced with the consequences of Francis Bacon's advocacy of science as power over nature, Goethe's 'delicate empiricism which makes itself utterly identical with the object' seems very attractive. This may well lead to a technology which is synergic rather than aggressive, in which nature and man co-operate to their mutual enhancement. There is no misanthropy in Goethe's approach. He has no room for the view that nature would be better off without man. On the contrary, the kind of technology which would be compatible with Goethean science is one in which nature becomes more fully itself through man's intervention. Nature and man belong together. The onlooker perspective of modern science sees man and nature as if they were entirely external to one another. This attitude hides the necessity in their belongingness, replacing it with what seems to be no more than an accidental conjunction. The Goethean way replaces this external dualism of man and nature with a deeper understanding of their relationship.

In view of the growing interest which is now evident in Goethe's approach to science, a carefully selected collection of Goethe's writings such as this one is much to be welcomed. It is very timely. We are all indebted to Jeremy Naydler for undertaking to collect these extracts, and for contributing such lucid and helpful introductions to the different sections into which he has organized them. It is a compendium which will surely be of use to anyone who wishes to look into the question of Goethe's science for themselves.

Henri Bortoft

Introduction

Today, one of the greatest obstacles facing people who are wanting to develop a more intimate relationship to nature is, paradoxically, the extraordinary 'success' of mainstream science. In the twentieth century, we have learned to distrust our own immediate experiences of nature, for we are educated into the belief that the world is not really as we experience it. Science has persuaded us that 'real reality' is a more or less abstract world behind the world we normally perceive — a world which lacks the qualities such as warmth, colour, taste, hardness or softness, by which we know and relate to nature in our ordinary lives. This world is accessible only to experts and specialists who have undergone the training necessary to see through the qualitative sheen that nature presents to us, to the substructure of essentially qualityless particles and processes which can only be adequately understood using mathematical concepts. It is a realm consisting of such things as protons, neutrons and electrons, quarks, leptons and probability waves, genes and DNA. The vast majority of people will never have even the remotest experience of them, for they can only be encountered under laboratory conditions, in a most indirect way. And even then only by people who have undergone the requisite training in order to interpret the phenomena they are observing. The rest of us have meanwhile fallen into the grip of a kind of paralysis regarding our relationship to nature. We no longer trust our own perceptions, and feel that to see the world 'as it really is' would require submitting ourselves to the awesome disciplines of the sciences. Convinced of the essential inferiority of our direct experience of nature, we have learnt to relinquish the challenge of deepening this experience in favour of turning to textbooks and encyclopedias in which the authorized definitions, classifications and interpretations of nature are preserved.

The advances in specialized scientific knowledge have thus had the unpredictable effect of encouraging a chronic laziness with regard to our everyday observation of nature. This applies not only to non-scientists, but equally to scientists themselves outside their particular area of specialism. There is now a widespread tendency to bring concepts to bear on our experiences in such a way that the experiences themselves are effectively undermined. This can happen either through our deferring to some underlying imperceptible agency described in the scientific textbook as the true cause of what we are experiencing, or by the conditioned reflex that makes us transpose our perceptions into something pre-defined and pre-classified. Before we allow ourselves the possibility of fully entering into them, and making them our own, we surrender our perceptions to a standard that is extraneous to our actual perceptual experience. In our Western scientific culture, we have reached a situation in which there is no longer a mental openness to phenomena, because we are so ready to defer to the pre-existent categories of established science.

The modern relinquishment by individuals of responsibility for their perception of nature has adversely affected the way in which our whole culture interacts with the natural world. Overwhelmed by the sheer quantity, complexity and brilliance of scientific knowledge, this interaction has become increasingly ignorant and insensitive. Here, then, is an aspect of the relationship between science and the contemporary ecological crisis which is critical, but is often overlooked. Unless we address it, any attempt to heal our wounded natural environment is unlikely to succeed. The re-sensitization of our day-to-day consciousness of nature has become a task that nature herself is pressing upon us. It is a task both more profound and more far-reaching than political legislation, economic strategy, or technological intervention. It is the key to restoring a harmonious relationship between ourselves and the natural world.

What Goethe had to say two centuries ago about nature and about science is, I believe, even more relevant now than it ever has been. For it bears directly on an issue which increasingly impinges on the lives of everyone: this issue is not simply the degradation of nature, but the degradation of our awareness of nature. Goethe was deeply concerned with science. But his concern was with a side of science which already, in his time, was being pushed to the periphery of what was generally regarded as having research value: namely, the intensification of our actual experience of the living world of qualities and forms, to a point at which one becomes aware of their spiritual, as distinct from their material, basis. For Goethe, this could be a perception as objective as the analysis of the chemical constituents of a body, but it required a training not only of the mind and senses but also of the imagination, and the moral and aesthetic sensibilities of the human being, all of which would be utilized by him in his researches. Science, for Goethe, was a method of consciousness-raising, but it could only function as such when brought into living connection with all his human faculties. None of these were excluded; rather, all were directed towards instilling vitality into the act of perception.

Goethe turned his attention to an extraordinarily wide range of natural phenomena. Mineralogy, geology and botany were all subjects of intense study in his late twenties when, having settled in Weimar at the invitation of Duke Karl August, his official duties included overseeing the local mines, and his official residence was a Gartenhaus in a park on the edge of town. It is typical of Goethe's personality that these circumstances of his Weimar life provoked in him a fascination respectively for rocks and plants. His interest in plants was further stimulated by his getting to know the nearby Thuringian forests, as well as by the herbalists whom he met in the woodland countryside.[1]

However, Goethe's first serious contribution to science was to be in the field of anatomy — a subject in which he had been keenly interested for many years. This was his discovery of the intermaxillary bone in the human jaw in 1784, the existence of which had hitherto been denied. For Goethe, it was a discovery of great significance, for it proved that there was a basic anatomical model which human beings shared with all other higher animals, thus validating his intuitive belief in the underlying unity of nature. At about the same time, Goethe was writing on granite, and pursuing his studies of plants. Goethe had a deep admiration for Linnaeus, and through the 1780's (while he was in his thirties) was constantly consulting his works. However, the type of minute analysis characteristic of Linnean botany was not to be Goethe's way. In 1790, he produced his classic study, The Metamorphosis of Plants, *in which he followed the development of the plant through archetypal stages of alternating contraction and expansion. Goethe was attempting to see beyond the individual, and beyond the species, to more fundamental processes of growth in which all plants share. This is not to say that his method was un-empirical; rather, he was striving for a kind of empiricism which attended to subtle processes and developmental patterns with which traditional taxonomy was little concerned.*

The following year, Goethe published his first essay on optics, a subject which was to preoccupy him for the rest of his life, and which led to his monumental study of colour phenomena, Zur Farbenlehre, *or* The Theory of Colour *(eventually published in its entirety in 1810). This work, more than any other of his scientific writings, has given him a place — a somewhat ambivalent place — in the history of science, and remains the subject of heated controversy. For it is a model of a type of scientific thinking utterly opposed to that of Newton, whose* Optics *dominated the field in Goethe's day. The essential difference in the methods of the two men is that Newton sought to explain the phenomenon of colour in terms of the measurable angles of refrangibility of colourless rays of light, whereas Goethe had no desire to reduce colour phenomena to what was measurable but colourless. His approach consisted in trying to understand colour in its own terms, and in terms of how we actually experience it as arising in nature.* Zur Farbenlehre *has always appealed more to artists who work with colour, than to scientists. But it also appeals to those unorthodox scientists and laypersons interested in developing a qualitative approach to nature, no less rigorous than the mathematical methods of the Newtonians and their descendants, but committed to maintaining faith with lived experience.*

Zur Farbenlehre *was not well received by the scientific establishment in Goethe's lifetime, and later physicists have often been less than sympathetic. In the latter half of the nineteenth century, Du*

Bois-Reymond, for instance, dismissed it as 'the still-born plaything of a self-taught dilettante,'[2] while Helmholtz, although he understood that Goethe was attempting to save the truth of direct sense-experience, nevertheless accused Goethe of failing to understand what constituted a scientific explanation. In his essay 'On the Natural-Scientific Works of Goethe,' Helmholtz was to comment tellingly:

> *Since we never are able to perceive the forces them-selves, but only their effects, in every explanation it is necessary for us to abandon the sensory realm and to pass over to the imperceptible, which is determined through concepts alone.[3]*

To judge from these reactions of establishment physicists, one suspects that Goethe, in his refusal to abandon the sensory realm in his pursuit of scientific knowledge, exercised a certain fascination over them. They could not ignore him, because he stood for a type of scientific investigation which they had rejected. They had rejected it in order to pursue what they knew was a one-sided knowledge of nature, and they felt it necessary to justify the one-sidedness of their approach by denouncing the man who represen-ted the other, neglected side of the scientific ideal. It was precisely Goethe's refusal to abandon the sensory realm that gave to his scientific endeavours a freshness and an accessibility that their science lacked, and indeed so much contemporary science lacks. Goethe was pursuing a type of knowledge that human beings can claim as their own, a type of knowledge that enriches rather than impoverishes human experience of nature.

As well as his work on colour, which occupied him from the 1790's onwards, Goethe continued to busy himself with studying and writing on osteology, comparative anatomy, geology, botany, zoology and, later, meteorology. He also wrote extensively on scientific methodology, the focus of the present volume. All of this work was by no means a merely peripheral hobby or pastime to Goethe. It was integral to his way of being in the world, a way of being in which he sought always to extend his embrace of nature through intensified experience and deepened understanding. To-wards the end of his life, Goethe wrote:

> *For more than half a century I have been known as a*
> *poet, in my own country and undoubtedly also abroad; or*
> *at any rate I have been permitted to pass for one. But the*
> *fact that I have busily and quietly occupied myself with*
> *Nature in all her general and organic phenomena,*
> *constantly and passionately pursuing seriously formulated*
> *studies — this is not so generally known, still less has it*
> *been accorded any attention.*[4]

Goethe the scientist is still little known, but it may be that the full significance of his contribution is now beginning to dawn on us.

Interestingly, with the collapse of the Newtonian paradigm in the twentieth century, Goethe's scientific work has received far more sympathetic attention from physicists than it did in the nineteenth century. Physicists such as Walter Heitler, Carl Friedrich von Weisäcker and Werner Heisenberg have all written on Goethe, Heisenberg in particular being aware of Goethe's voice almost as the troubled conscience of the modern scientist.[5] *In his essay 'On the History of the Physical Interpretation of Nature,' Heisenberg writes of the basis of scientific progress since Newton's time as having involved a sacrifice of 'living and immediate under-standing,' and accepts that this 'was the real reason for Goethe's bitter struggle against Newton's physical optics and his teachings on colour.' He continues:*

> *It would be superficial to neglect this struggle as*
> *unimportant, there is a good reason for one of the most*
> *eminent of men using all his power to combat the*
> *achievement of Newton's optics. One can only charge*
> *Goethe with a lack of consistency. He should not only*
> *have combated Newton's views but he should have said*
> *that the whole of Newton's physics, optics, mechanics and*
> *gravitational theory was the work of the devil.*[6]

For Heisenberg, there need be no conflict between accepting the findings of modern physics and 'following Goethe's way of contemplating nature.'[7] For the two ways are less opposed than complementary.

Within the world of biology, Goethe's reception in the twentieth century has generally lacked this degree of broadmindedness. Sir Charles Sherrington, in his essay 'Goethe on Nature and on Science' written in the 1940's (just a few years after Heisenberg was writing his thoughts on Goethe), gives a fairly typical verdict on Goethe's Metamorphosis of Plants:

> *Now the study of the development of a plant, or animal, is at root an affair of following its cellular development. The cell-theory had not arrived in Goethe's time ... When later the progress of botany in due course obtained facts competent for the question, the theory was found not to be borne out. It fell, therefore, into the doleful category of unlucky guesses.[8]*

This has not stopped some more recent biologists, opposed to the reductionist tendency endemic in modern biology, to come out openly in support of Goethe. Notable amongst these are Agnes Arber, Adolf Portmann and Brian Goodwin.[9] It is not Goethe's specific discoveries or 'theories' that draw them to him, but rather the whole manner in which he conducted his investigations. Above all, it is Goethe's methodology that remains his lasting legacy to us. And it is precisely over his methodology that scientists either feel compelled to dismiss him as a poet who made a fool of himself dabbling in matters beyond his competence, or find themselves irresistibly attracted to him as a pioneer of a holistic and qualitative science of nature. It is worth quoting Portmann here, for he expresses succinctly the inspirational role that Goethe can play today. At the end of his essay on 'Goethe and the Concept of Metamorphosis,' Portmann writes:

> *It is high time we rediscovered the exemplary nature of an attempt such as that which Goethe has given us in his*

Metamorphosis of Plants ... *The accelerated development of biological research in the direction of genetic engineering, that investigates the visible realm in order to achieve mastery over the processes of nature — this unavoidable development will result in a horrifying impoverishment of our relationship to nature if we do not begin immediately to take to heart the value of an extensive experience with living form for the cultivation of the soul. New forms of science of nature are called for, a science of nature which is not a pale reflection of today's science, but rather leads to a deepened experience with the realm of living forms and makes nature for us a true home.*[10]

It was Goethe's commitment to the revitalization of our perception of the world so that we again find ourselves at home within nature instead of studying her as if we were aliens from another planet that is his major contribution to science and, more broadly, to our beleaguered scientific culture today. And yet, for non-German readers, there have been few opportunities to come to know Goethe's thoughts at first hand. For a long time, the principal source of his writings on scientific method in English was Bertha Mueller's selection in Goethe's Botanical Writings *(1952).*[11] *This volume contains translations of all Goethe's important essays on scientific methodology, but it was never reprinted and has been virtually unobtainable for decades. Most of these essays, together with some previously untranslated writings by Goethe, have now been retranslated in Douglas Miller's* Goethe: Scientific Studies *(1988).*[12] *These two volumes are key sources for much that appears in the present collection.*

Apart from these two important anthologies of Goethe's writings on science, there have been numerous attempts to present Goethe's outlook in a more or less systematic way. The most successful of these attempts in recent years is undoubtedly Henri Bortoft's Goethe's Scientific Consciousness *(1986).*[13] *Where the present collection differs from previous anthologies and expositions is that it is made with the intention of expounding in a* systematic *way, and* in Goethe's own words, *his central ideas on nature, science and scientific method. The present collection deliberately omits Goethe's studies of specific subjects, such as his study of colour, the metamorphosis of plants, his writings on zoology, geology, meteorology, etc. in order to focus on his unique conception of what the scientific approach to nature should be.*

As conceived by Goethe, science is as much an inner path of spiritual development as it is a discipline aimed at accumulating knowledge of the physical world. Rather than simply making new discoveries and propounding new theories on the basis of ever more refined techniques of physical observation, the aim of science is, for Goethe, to open the eyes and mind of the beholder of nature to what is spiritually at work within, or at root of, the observed physical phenomena. It therefore involves not only a rigorous training of our faculties of observation and thinking, but also of other human faculties which can attune us to the spiritual dimension that underlies and interpenetrates the physical: faculties such as feeling, imagination and intuition. Science, as Goethe conceived and practised it, has as its highest goal the arousal of the feeling of wonder through 'contemplative looking' (Anschauung), *in which the scientist would come 'to see God in nature, nature in God.'*[14]

Such an experience does not depend on the observer having accumulated vast amounts of knowledge. Nor does it rest on the deployment of elaborate scientific instruments. For Goethe, the human being *is 'the most powerful and exact instrument' if we but take the trouble sufficiently to refine our sensibilities.*[15] *This means that the practice of Goethean science is not necessarily the province of specialists and experts, but is open to everyone who is seeking to deepen their relationship to nature.*

From a Goethean standpoint, the ecological crisis is above all a crisis of our relationship to nature. The extent to which nature is in need of being healed corresponds directly to the extent to which our consciousness of nature is sick. The Goethean approach to nature rests upon the development of human consciousness towards a more wholesome perception of, and ever more subtle degrees of attunement to, the creative and formative forces within nature. Goethe shows us a path in which the healing both of nature and

*ourselves is implied in the 'delicate empiricism' which he es-
pouses.[16] It is essentially a reverential path, not a path of mani-
pulation and control. It is a path in which human beings may
become whole again in an experience of nature which opens to the
sacred, and which thus becomes the means by which nature herself
is re-sanctified. Above all, Goethe's scientific path is a path which
keeps faith with human experience, and seeks less to move from
experience to idea or theory, than to intensify experience as such.
It is through this intensification of our experience of nature that
her spiritual dimension is revealed.*

REFERENCES

1. See Goethe's essay, 'The Author Relates the History of His Botanical Studies'
   trans. Mueller, pp.150–51 (see note 11).
2. Du Bois-Reymond, 'Goethe ad infinitum' quoted in Fred Amrine, 'Goethe's
   Science in the Twentieth Century,' in Alexej Ugrinsky, *Goethe in the Twentieth
   Century* (Hofstra University, 1987) p.88.
3. H. von Helmholtz, 'On the Natural-Scientific Works of Goethe' quoted in Fred
   Amrine, *op. cit.* p.88.
4. 'The Author Relates the History of His Botanical Studies' trans. Mueller p.164.
5. See Walter Heitler, *Man and Science* (trans. R. Schapp, New York, Basic
   Books, 1963), Carl Friedrich von Weisäcker, 'Goethe and Modern Science' in
   Fred Amrine (ed.), *Goethe and the Sciences: a Reappraisal* (Dordrecht,
   D. Reidel Publishing Co., 1987), and Werner Heisenberg, *Across the Frontiers,*
   New York, Harper and Row, 1974).
6. Werner Heisenberg, 'On the History of the Physical Interpretation of Nature'
   in *Philosophic Problems of Nuclear Science* (London, Faber and Faber, 1952),
   reissued as *Philosophic Problems of Quantum Physics* (Woodbridge,
   Connecticut, Ox Bow Press, 1979).
7. Werner Heisenberg, 'The Teachings of Goethe and Newton on Colour in the
   Light of Modern Physics' in *Philosophic Problems of Nuclear Science,* p.75.
8. Sir Charles Sherrington, *Goethe on Nature and on Science,* (Cambridge
   University Press, 1949) p.22.
9. See Agnes Arber, *The Natural History of Plant Form,* (Cambridge University
   Press, 1950) and her introduction to 'Goethe's Botany' in *Chronica Botanica,*
   Vol.X, No.2, 1946; Adolf Portmann, *New Paths in Biology,* (New York, Harper
   and Row, 1964) and his essay 'Goethe and the Concept of Metamorphosis' in
   Fred Amrine, *Goethe and the Sciences: a Reappraisal (op.cit.);* Brian Goodwin,
   'Towards a Science of Qualities' in W. Harman and J. Clark (eds.), *New
   Metaphysical Foundations of Modern Science,* (Sausalito, Institute of Noetic

Sciences, 1994), and *How the Leopard Changed its Spots,* (London, Weidenfeld and Nicholson, 1994).

10. Adolf Portmann, 'Goethe and the Concept of Metamorphosis' in Fred Amrine, *Goethe and the Scientists: a Reappraisal (op.cit.),* pp.144–45.

11. Bertha Mueller, *Goethe's Botanical Writings,* (Honolulu, University of Hawaii Press, 1952).

12. Douglas Miller, *Goethe: Scientific Studies,* (New York, Suhrkamp, 1988, New edition, Princeton University Press, 1995).

13. Henri Bortoft, *Goethe's Scientific Consciousness,* (London, Institute of Germanic Studies, 1972). This is reproduced in Henri Bortoft's fuller study, *The Wholeness of Nature: Goethe's Way of Science,* (Edinburgh, Floris Books 1996; New York, Lindisfarne, 1996).

14. Chapter 8, §1.

15. Chapter 1, §1.

16. See Chapter 5.

1. The Human Being is the most Exact Instrument

The proposition that the human being is the most exact scientific instrument contradicts what most contemporary scientists take for granted, and runs counter to the whole way in which science has been practised in modern times. Without the development of ever more refined and sophisticated non-human instruments, most of the advances in modern scientific knowledge simply would not have occurred. It therefore seems contrary to common sense to claim that the human being is the most exact instrument, when there are instruments far more sensitive than any human being, and it is upon these instruments that the modern scientist depends. Even in Goethe's day this was the case. Why, then, should Goethe insist that the human being is the most exact instrument?

The answer lies in Goethe's very different conception of what it is that scientists should ultimately be directing their attention towards. Since the sixteenth and seventeenth centuries, when the foundations of the modern scientific enterprise were laid, it has been an almost unquestioned assumption within mainstream science that all experienced qualities and forms in nature are caused by further microscopic qualities not directly experienced, but made accessible through various instruments and analytical techniques. These further phenomena, unlike those we meet in our normal experience, are susceptible to exact chemical and mathematical analysis, and should be regarded as the 'primary' qualities, or 'nature's building blocks,' whereas the qualities we directly experience in the normal course of our lives, are rather inexact 'secondary' qualities. Secondary qualities are but epi-phenomena of this underlying reality, and can be wholly explained in terms of the primary qualities of which they are the by-product.

While Goethe could see the benefits of ever more precise observation through the use of instruments, he held that unless these observations were brought back into connection with the lived human relationship to nature, they would inevitably lead to a one-sided and distorted understanding of the world. Things would be explained in such a way that their reality would be denied through the very terms by which they were explained. Goethe had experienced this already in the Newtonian explanations of colour-phenomena, which reduced colours ('secondary' qualities) to no more than angles of refraction of colourless rays of light ('primary' qualities). In a similar way, modern biology has come to view living organisms as little more than the epi-phenomena of groups of genes — the 'DNA replicators' — which have a micro-scopic existence inside them. As a consequence, the organism as a living whole disappears from view. Today, this type of explanation has become so much a part of our way of understanding nature that any other kind of explanation is generally regarded as 'unscientific.'

But for Goethe, science should attend to wholes as well as parts, to directly experienced qualities as much as indirectly experienced microscopic ones. Such attentiveness does not depend on extending one's observations into a 'sub-phenomenal,' and also increasingly conceptualized, world that forms no part of our lived experience of nature. Goethe was convinced that a contemplative observation of qualities and forms in nature encountered directly, even naively, is sufficient basis for an awareness of the non-material formative forces and organizational principles which underlie them to arise. This other, spiritual side of the phenomenal world cannot be penetrated by instruments sensitive only to what is material, nor by the more or less abstract interpretive concepts which accompany their use. This is why the human being, contemplating natural phenomena with alert senses and an open mind, is potentially a more powerful and exact instrument than any piece of specialized scientific equipment.

~ *1* ~

(a) In so far as we make use of our healthy senses, the human being is the most powerful and exact scientific instrument possible. The greatest misfortune of modern physics is that its experiments have, as it were, been set apart from the human being; physics refuses to recognize in Nature anything not shown by artificial instruments, and even uses this as a measure of its accomplishments.

(b) The same applies to calculation. Many things cannot be calculated, and there are other things which defy experiments.

(c) But in this connection the human being stands so high that what otherwise defies portrayal is portrayed in us. What is a string and all mechanical subdivisions of it compared with the ear of the musician? Yes, indeed, what are the elemental phenomena of Nature herself in comparison with the human being, who must first master and modify them in order in some degree to assimilate them?

~ *2* ~

Next, you must trust your senses:
they will show you nothing false
if your intelligence keeps you awake.

Keep your eyes fresh and open and joyful,
and move with sure steps, yet flexibly,
through the fields of a world so richly endowed.

Den Sinnen hast du dann zu trauen,
Kein Falsches lassen sie dich schauen,
Wenn dein Verstand dich wach erhält,

Mit frischen Blick bemerke freudig.
Und wandle, sicher wie geschmeidig.
Durch Auen reichbegabter Welt.

~ *3* ~

The human being is adequately equipped for all genuine needs on earth if we trust our senses, and develop them in such a way that they continue to prove worthy of confidence.

~ *4* ~

The senses do not deceive; the judgment deceives.

~ *5* ~

Microscopes and telescopes in fact confuse pure common sense.

~ *6* ~

People who look through glasses [that is, microscopes and telescopes] think themselves cleverer than they are: for their external sense is in this way taken out of equilibrium with their inner capacity for judgment.

~ 7 ~

Someday someone will write a pathology of experimental physics and bring to light all those swindles which subvert our reason, beguile our judgment and, what is worse, stand in the way of any practical progress. The phenomena must be freed once and for all from their grim torture chamber of empiricism, mechanism, and dogmatism; they must be brought before the jury of man's common sense.

~ 8 ~

Few people have the gift of grasping nature and using it directly; between knowledge and application they prefer to invent a phantom which they develop in great detail; doing so, they forget both object and purpose.

~ 9 ~

The Greeks spoke of neither cause nor effect in their descriptions and stories — instead, they presented the phenomenon as it was. In their science, too, they did not perform experiments, but relied on experiences as they occurred.

~ 10 ~

The animal is instructed by his sensory organs; man instructs his organs and governs them.

~ *11* ~

This is what they all come to who exclusively harp on experience. They do not stop to consider that experience is only one half of experience.

~ *12* ~

In the sciences everything depends on what one calls an *aperçu* — the discovery of something that is at the bottom of phenomena — such a discovery is infinitely fruitful.

~ *13* ~

The desire for knowledge first stirs in us when we become aware of significant phenomena which require our attention. To sustain this interest we must deepen our involvement in the objects of our attention and gradually become better acquainted with them. Only then will we notice all manner of things crowding in upon us. We will be compelled to distinguish, differentiate and resynthesize, a process which finally leads to an order we can survey with some degree of satisfaction.

To achieve this even partially in any field of knowledge requires constant and rigorous effort. Therefore we find that many would prefer to dismiss phenomena with a general theoretical precept or a quick explanation without taking the trouble to study them in detail and achieve a knowledge of the whole over a longer period of time.

~ *14* ~

The investigator of Nature should take heed not to reduce observation to mere notion, to substitute words for this notion, and to use and deal with these words as if they were things.

~ *15* ~

How difficult it is, though, to refrain from replacing the thing with its sign, to keep the object alive before us instead of killing it with a word.

~ *16* ~

In reality, any attempt to express the inner nature of a thing is fruitless. What we perceive are effects, and a complete record of these effects ought to encompass this inner nature. We labour in vain to describe a person's character, but when we draw together their actions, their deeds, a picture of their character will emerge.

~ *17* ~

Experiencing, looking, observing, contemplating, connecting, discovering, inventing are mental activities which, singly and severally, are exercised a thousandfold by more or less gifted people ... From these various powers named here, and many other related ones, Mother Nature has excluded no one.

~ *18* ~

Many people may be persuaded to make observations for themselves, within their own immediate sphere. And such individual
observations, drawn from the natural objects with which we are in
constant contact, are often the more valuable the less the observer
professionally belongs to a particular department of science ... . As
soon as anybody belongs to a certain narrow creed in science,
every unprejudiced observation is gone ... . The contemplation of
the world, with all these theorists, has lost its innocence, the
objects no longer appear in their natural purity. If these learned
people, then, give an account of their observations, we obtain, not-
withstanding their love of truth as individuals, no actual truth with
reference to the objects; we always get the taste of a strong subjective mixture.

I am, however, far from maintaining that an unprejudiced correct
knowledge is a drawback to observation. I am much more inclined
to support the old truth, that we have only eyes and ears for what
we know. The professional musician hears, in an orchestral per-
formance, every instrument and every single tone; whilst one
unacquainted with the art is wrapped up in the massive effect of
the whole. Someone merely bent upon enjoyment sees in a green
or flowery meadow only a pleasant view, whilst the eye of a botan-
ist discovers endless detail of the most varied plants and grasses.

All have their measure and goal; and, as it has been said in my
*Goetz von Berlichingen,** that the son, from pure learning, does not
know his own father, so in science do we find people who can
neither see nor hear, through sheer learning and hypothesis. Such
people look at once within; they are so occupied by what is
revolving in themselves that they are like someone in a passion,
who passes their closest friends in the street without seeing them.
The observation of Nature requires a certain purity of spirit that
cannot be disturbed or preoccupied by anything. The beetle on the
flower does not escape the child; he has devoted all his senses to

---

* Goethe's dramatic chronicle of the life of a sixteenth century German baron,
  published in 1773.

a single simple interest; and it never strikes him that at the same moment something remarkable may be going on in the formation of the clouds to distract his glances in that direction.

Would to God we were all nothing more than good hod-carriers. It is just because we will be more, and carry about with us a great apparatus of philosophy and hypothesis that we spoil everything.

2. Observation of Nature is Limitless

For Goethe, nature's complexity and subtlety demands from us an inner flexibility and openness to her manifold phenomena that would be impossible were we to remain rigidly within just one mode of awareness. The scientist sees one aspect of nature, the metaphysician another, the poet yet another. Within the sciences, too, a given phenomenon can be apprehended from a variety of different standpoints, and from within a range of different but complementary disciplines. Each mode of human observation is sensitive to but one dimension of nature's multi-dimensional existence.

A comprehensive grasp of nature will therefore require of us that we explore many different types of knowing, and avoid restricting ourselves to one single style of knowledge. The Goethean attitude is the opposite of the reductionist tendency that would give validity to just one way of viewing nature, and deny validity to all others. For Goethe, the wider the range of viewpoints we adopt, the more comprehensive will our perception of nature become, and hence the more complete our understanding will be.

~ *1* ~

Natural system: a contradictory expression.

Nature has no system; she has — she is — life and development from an unknown centre toward an unknowable periphery. Thus observation of Nature is limitless, whether we make distinctions among the least particles or pursue the whole by following the trail far and wide.

~ *2* ~

If you want to reach the infinite,
explore every aspect of the finite.

Willst du ins Unendliche schreiten,
Geh nur im Endlichen nach allen Seiten.

~ *3* ~

Central to Goethe's view of science is his belief that there is no
exclusively 'right' approach to nature. Different scientists may
legitimately adopt the approach most suited to their own funda-
mental disposition, be they utilitarians, fact-finders, contemplators
or 'comprehenders.' The distinction between them lies in the
degree to which they utilize the more internal human faculties of
thinking, imagination and creative intuition. To the extent that they
do, they are exposed to the danger of subjective projection; on the
other hand, if this danger is overcome, the more 'internal' ap-
proaches promise a deeper understanding of, and participation in,
nature's creative processes. The following is an extract from notes
written by Goethe in the 1790's, published posthumously in the
Weimar edition of his works.

... no one asks a question of Nature that they cannot themselves
answer, for the answer is inherent in the question, in the feeling
that the point can be discussed and pondered.

To be sure, the questions vary according to the different types of
humans.

To orient ourselves somewhat among these various types, let us
divide them thus into four spheres: utilizers, fact-finders, contemplators, and comprehenders.

(1) The utilizers, advocates and seekers of things practical, are the
first to plough the field of science, metaphorically speaking, and

they aim at practical results. Self-confidence derived from experience gives them assurance; necessity gives them a certain breadth.

(2) Fact-finders, those who crave knowledge for its own sake, require a calm, disinterested gaze, an inquisitive unrest, a clear mind. They are in contrast with the first group, but work out the results from the scientific point of view exclusively.

(3) The contemplators are somewhat more original, for the mere increase of knowledge unwittingly fosters interpretation and crosses over into it. Even the fact-finders, however much they may make the sign of the crucifix at the very thought of imagination, before they realize it are compelled to call upon this selfsame power for assistance.

(4) The comprehenders — in a deeper sense they might be called creators — are original in the highest sense of the term. By proceeding from ideas, they simultaneously express the unity of the whole, and it is almost the obligation of Nature to conform to the ideas.

Simile of roads.
   Illustration of the aqueduct, to distinguish between the fantastic and the ideal.
   Illustration of the dramatic poet.
   Creative imagination with all possible realism.
   In all scientific effort one must make clear to oneself that one will move in these four spheres.
   One must ever be conscious of the sphere one is working in at the moment.
   And one must have the inclination to move in one as freely and easily as in another.
   The objective and subjective in one's exposition is thus recognized and separated in advance, and in this way one can at least hope to inspire a degree of confidence.

~ *4* ~

In these paragraphs from the Theory of Colour, *Goethe struggles with the problem of finding a language adequate to the phenomena observed in nature. Whether one uses metaphysical or mathematical language, mechanical or moral language, in each case one comes up against the limitations intrinsic to a particular way of relating to nature. Ideally, scientists will utilize many different languages, because their approach to nature will be as comprehensive and as all-embracing as possible.*

We are insufficiently aware that a language is, in fact, merely symbolic, merely figurative, never a direct expression of the objective world, but only a reflection of it. This is especially so when we speak of things which only touch lightly upon our empirical observation, things we might call activities rather than objects. In the realm of natural philosophy such things are in constant motion. They cannot be held fast and yet we must speak of them; hence we look for all sorts of formulas to get at them, at least metaphorically.

Metaphysical formulas have great breadth and depth, but a rich content is required to fill them in a worthy way; otherwise they remain empty. Mathematical formulas are often convenient and useful, but they always have a certain stiffness and awkwardness; we soon feel their inadequacy, for even in elementary instances we will quickly recognize the presence of an incommensurable quality. Furthermore, they are intelligible only to a narrow circle of specially trained minds. Mechanical formulas speak more to ordinary understanding, but are themselves ordinary and always retain a touch of crudity. They transform living things into dead ones; they kill the inner life in order to apply an inadequate substitute from without. Corpuscular formulas are similar; they have the effect of rigidifying things in motion, coarsening idea and expression. In contrast, moral formulas express more delicate relationships but take the form of simple metaphors, and may finally lose themselves in a display of wit.

However, scientists might make conscious use of all these modes

of thought and expression to convey their views on natural phenomena in a multifold language. If they could avoid becoming one-
sided, and give living expression to living thought, it might be
possible to communicate much that would be welcome.

How difficult it is, though, to refrain from replacing the thing
with its sign, to keep the object alive before us instead of killing
it with the word. In recent times this danger has been heightened
as expressions and terms are drawn from all areas of knowledge
and science to express perceptions of simple natural phenomena.
We call on the aid of astronomy, cosmology, geology, natural
history, even religion and mysticism; and often the particular, the
derived, will hide and obscure the general, the elementary, instead
of illuminating and revealing it. We are quite aware of the
necessity responsible for such a language and its widespread use,
and we know that it has made itself indispensable in a certain
sense. But this language will be of service only when more
moderately and modestly applied in a conscious and sure way.

~ 5 ~

In New York there are ninety different Christian sects, each
acknowledging God and our Lord in its own way without
interference. In scientific research — indeed, in any kind of
research — we need to reach this goal; for how can it be that
everyone demands open-mindedness while denying others their
own way of thinking and expressing themselves?

~ 6 ~

In the sciences ... a continual circulation takes place — not because
the objects themselves change, but because new observations
produce a need in each scientist to assert himself, to handle knowledge and the sciences in his own way.

But since human thought also follows a certain circular pattern, a reversal of method will always bring us back to the same point. These atomistic and dynamic concepts will forever alternate, but only in emphasis, for neither will wholly replace the other. This holds true even for the individual scientist. Before he realizes it, the most determined dynamist will fall into atomistic terminology, while the atomist will be unable to avoid becoming dynamistic at times.

~ 7 ~

In this extract from his 'Preliminary Notes for a Physiology of Plants,' Goethe outlines the spheres of operation of the various sciences concerned with the study of living organisms. Goethe was not opposed to the orthodox scientific approach to nature, but sought only to extend it towards a more holistic perception of phenomena. This would involve both the integration of the various disciplines and their subsumption under the overarching science of morphology (that is, the study of forms).

In an organic being, first the form as a whole strikes us, then its parts and their shape and combination.

Form in general and the relation and combination of the parts, in so far as they are outwardly visible, constitute the scope of natural history. However, when these parts are first presented to the eye in isolated form, we are concerned with the science of anatomy. The latter does not deal merely with outward form but with inner structure and thus quite properly makes use of the microscope.

When the organic body has been broken down in this way, so that its form is dissolved and its parts can be regarded as matter, the science of chemistry sooner or later steps in to give us new and wonderful information concerning the smallest ultimate parts and their combination.

When the destroyed creature is recreated from all these singly noted phenomena and is observed in its live and healthy state, we call the activity physiology.

Physiology is the mental operation performed in attempting to put together a whole from the animate and the inanimate, the known and the unknown, the complete and the incomplete, from perceptions and conclusions. Such a whole is simultaneously visible and invisible; its exterior must forever remain a whole to us, its interior forever a part; its actions and effects must remain eternally a mystery. Thus is it easy to see why physiology had to lag behind for so long and probably always will lag behind, namely, because human beings, though they feel their limitations, are seldom willing to acknowledge them.

Anatomy has been elevated to such a degree of exactness and precision that its well-defined facts in themselves comprise a kind of physiology.

Bodies are moved in so far as they have length, breadth, and weight, in so far as pressure and thrust act upon them, and in so far as they can be moved in some manner or other. For that reason people familiar with the laws of Nature have applied them also to organic bodies and their movements, not without advantage.

Chemists, too, have observed in detail the changes as well as the composition of the smallest ultimate parts, and their diligence of late as well as the accuracy of their methods justify their claims of having revealed the nature of the organic complex.

On the basis of all this, even aside from considerations not taken into account here, one can easily see how necessary it is to summon all our mental powers in a general striving for insight into all these mysteries, how necessary it is to use all mental and physical aids and to press every advantage in venturing to approach this never-ending work. Even a certain one-sidedness is not disadvantageous to the whole. Indeed, let each individual consider their own way best, if only they will smooth and clear it properly, so that those who follow them on the road may traverse it more easily and rapidly.

Recapitulation of the various sciences:

(a) Knowledge of organic bodies, according to habitat and differences in form relationships.
  Natural history.

(b) Knowledge of material natures in general, as forces and in their place relationships.
  Natural philosophy.

(c) Knowledge of organic bodies, with reference to their inner and outer parts, without consideration of the living whole.
  Anatomy.

(d) Knowledge of the parts of an organic body in so far as it ceases to be organic, or in so far as its organization is regarded merely as substance-producing and substance-composed.
  Chemistry.

(e) Observation of the whole in so far as it is living and its life has a special physical power.
  Zoonomy.

(f) Consideration of the whole in so far as it lives and acts, and in so far as an immaterial power is at the basis of this life.
  Physiology.

(g) Consideration of form both in its parts and as a whole, the conformities and deviations, apart from all other consideration.
  Morphology.

(h) Consideration of the organic whole by visualizing and linking all these considerations through mental processes.

~ *8* ~

Anatomy does for organized beings what chemistry does for unorganized matter.

~ *9* ~

To escape the endless profusion, fragmentation, and complication of modern science and recover the element of simplicity, we must always ask ourselves: what approach would Plato have taken to a nature which is both simple in essence and manifold in appearance?

~ *10* ~

Four epochs of science:

childlike,
poetic, superstitious;

empirical,
searching, curious;

dogmatic,
didactic, pedantic;

ideal,
methodical, mystical.

~ *11* ~

(a) A crisis must necessarily arise when a field of knowledge matures enough to become a science, for those who focus on details and treat them as separate will be set against those who have their eye on the universal and try to fit the particular into it. Now, however, an ideal, more comprehensive scientific approach is attracting an ever wider circle of friends, patrons, and colleagues; at this higher stage the division is no longer so marked, although still noticeable enough.

Those I would call 'universalists' hold firm to the conviction that everything is present everywhere and may be discovered there, although in forms endlessly divergent and varied. The others, whom I will call 'singularists,' agree with this principle, and even follow it in their observations, definitions, and teachings. But they claim to find exceptions wherever the prototype is not fully expressed, and rightly so. Their only error lies in failing to recognize the basic form where it is disguised, and denying it where it is hidden. Yet both ways of thought are authentic. They stand in eternal opposition with no prospect of joining forces or defeating one another: hence we must avoid engaging in controversy and simply state our convictions clearly and openly.

(b) I will therefore restate mine: at this higher level we cannot *know,* but must *act,* just as we need little knowledge but much skill in a game. Nature has given us the chess board; we cannot and should not work beyond its limits. She has carved our pieces; gradually we will learn their value, their moves, and their powers. Now it will be our task to find the moves we think best; each seeks this in their own way regardless of any advice. Leave well enough alone, then. Let us merely observe the distance between us and the others, finding our allies in those who declare themselves on our side. We should also recall that we are dealing with an insoluble problem. We must be ready to attend to anything we may hear, especially anything opposed to our own view, for here we will recognize the problematic character of things and, especially, of people. I am not sure I will continue my work in this well-tilled field, but I reserve my right to note and point out certain new directions of study or individual research.

~ *12* ~

I, for my part, drawn in many directions as I am, cannot content myself with one way of thinking. As poet and artist I am a polytheist; in my nature studies I am a pantheist — both in a very determined way. When I require one god for my personality, as an ethical being, this is provided for also. The things of heaven and earth contain such a wealth of value that only the organs of all beings jointly can encompass it.

~ *13* ~

The worthiest professor of physics would be one who could show the inadequacy of his text and diagrams in comparison to nature and the higher demands of the spirit.

3. Setting Forth a Morphology

In his 'Preliminary Notes for a Physiology of Plants' (see p.41ff), Goethe specifically advocates a physiology based on an improved understanding of the physico-chemical basis of life. But he also points out that while living beings make use of physico-chemical substances in achieving their ends, they cannot simply be reduced to these substances. It is necessary also to consider 'the whole in so far as it lives and acts and in so far as an immaterial power is at the basis of this life' (p.41).

In the late twentieth century, when there is so much emphasis within biology on molecular research, microbiology and biochemistry, the idea of 'an immaterial power' as the basis of the life of an organism must appear to many as an unnecessary metaphysical intrusion. For Goethe, however, if we are to perceive the living development of organisms as having an overall coherence, then we are directed towards an organizational principle that cannot be apprehended by the kind of minute analysis that characterizes modern research in biology. Goethe's 'immaterial power' is the morphotype, which is neither reducible to the constituent physical parts of an organism, nor can it be identified with any one particular stage of an organism's development, no matter how apparently 'archetypal' this stage may seem. For the morphotype is both what organizes the constituent parts of an organism into a harmoniously functioning unity, and what guides an organism's development so that its varied manifestations in time are all expressions of this same underlying unity. Morphology is the scientific study in which due respect is paid to the functional relationship of both spatial and temporal aspects of an organism within the whole, but in which the emphasis is firmly placed on the relationship of these aspects to the whole, which cannot be identified with any one of them.

Literally 'the study of forms,' morphology was for Goethe the most universal and hence the most important of the sciences. Since Goethe himself coined the word 'morphology,' he has been duly recognized as the founder of modern comparative morphology. As Goethe understood it, however, morphology should aim less to study completed *forms, than the inner formative forces which give rise to them. Through close observation of physical structures and processes, it should prove possible to arrive at a more interior perception of the form-making power of which they are a manifestation. In order to reach such an interior perception, it is necessary to shift one's consciousness from the analytical mode (analysing and dissecting physical and physico-chemical features of an object) to a wholistic or synthesizing mode (grasping the living unity and coherence of an object as it develops in space and time). This shift in consciousness entails no longer seeing the object simply as an object but also as a spiritual subject. It is thus to glimpse the workings of the divine within nature.*

Morphology is therefore not only a science but it is simultaneously a spiritual path, by which the scientist activates a higher, more refined mode of thinking and observation than is normally utilized in general scientific practice. Goethe points towards the altered consciousness that is required by employing a distinction frequently made by the idealist philosophers of his day, between the Verstand *(ordinary analytical understanding) and the* Vernunft *(higher intuitive and synthesizing insight). The science of morphology is founded upon the activation of the* Vernunft, *which alone lays hold of nature as living, dynamic and creative.*

~ *1* ~

This extract is from an essay written by Goethe in 1807 (though not published until 1817) in which he outlines his approach to the study of living organisms. The essay was intended as an introduction to his botanical writings, and in the following passage Goethe states a fundamental principle of his approach to living organisms. He makes an important distinction between that which is already formed and thus fixed in character (Gestalt) *and the formative process as such* (Bildung). *He introduces the term 'morphology' (study of form) to describe his approach, which is characterized by a striving to reach beyond the fixed* Gestalt *to the dynamic* Bildung *which constantly creates or builds up new forms (the word* Bildung *is derived from the verb* bilden *to build).*

Nowhere does Goethe express so clearly the difference between his approach to the study of living organisms, and that of orthodox biology. While he accepts the valuable contribution of biochemistry to this study, and he would undoubtedly have acclaimed the spectacular advances in microbiology during the twentieth century, Goethe did not believe that the form-creating capacity of a living organism could be located simply by physico-chemical analysis. For Goethe, what is sense-perceptible — even if only indirectly with the aid of sophisticated laboratory equipment — is already formed. For this reason, though it may help the scientist to look towards what is formative, it cannot be equated with it. For what is formative cannot be identified with anything physical: its manner of existence is essentially creative. Hence, it is necessary to employ another kind of faculty, which Goethe calls Anschauung *(here translated as 'intuitive perception'), in order to see through the already formed, outwardly perceptible* Gestalten *to the underlying formative principle that welds them into a coherent and living whole.*

In observing objects of Nature, especially those that are alive, we often think the best way of gaining insight into the relationship between their inner nature and the effects they produce is to divide them into their constituent parts. Such an approach may, in fact, bring us a long way toward our goal. In a word, those familiar with science can recall what chemistry and anatomy have contributed toward an understanding and overview of Nature.

But these attempts at division also produce many adverse effects when carried to an extreme. To be sure, what is alive can be dissected into its component parts, but from these parts it will be impossible to restore it and bring it back to life. This is true even of many inorganic substances, to say nothing of things organic in nature.

Thus scientific minds of every epoch have also exhibited an urge to understand living formations as such, to grasp their outward, visible, tangible parts in context, to see these parts as an indication of what lies within and thereby gain some understanding of the whole through an exercise of intuitive perception *(Anschauung)*. It is no doubt unnecessary to describe in detail the close relationship between this scientific desire and our need for art and imitation.

Thus the history of art, knowledge, and science has produced many attempts to establish and develop a theory which we will call 'morphology.' The historical part of our discourse will deal with the different forms in which these attempts have appeared.

The Germans have a word for the complex of existence presented by a physical organism: *Gestalt* (structured form). With this expression they exclude what is changeable and assume that an interrelated whole is identified, defined, and fixed in character.

But if we look at all these *Gestalten,* especially the organic ones, we will discover that nothing in them is permanent, nothing is at rest or defined — everything is in a flux of continual motion. This is why German frequently and fittingly makes use of the word *Bildung* (formation) to describe the end product and what is in process of production as well.

Thus in setting forth a morphology we should not speak of *Gestalt,* or if we use the term we should at least do so only in reference to the idea, the concept, or to an empirical element held fast for a mere moment of time.

When something has acquired a form it metamorphoses immediately to a new one. If we wish to arrive at some living perception of Nature we ourselves must remain as quick and flexible as Nature and follow the example she gives.

~ *2* ~

The time will come when intelligent students will have discarded mechanistic and atomistic conceptions altogether in favour of viewing all phenomena in terms of dynamic and chemical processes, thus making the divine life in Nature more and more manifest.

~ *3* ~

The power of Godhead permeates what is living, not what is dead; it is present in that which is in process of becoming and that which transforms itself, not in that which has become and has congealed in its form. Hence reason *(Vernunft),* in its affinity with the divine principle, is concerned with what is evolving and living, whereas the understanding *(Verstand)* deals with what has become formed and congealed, in order to put it to use.

~ *4* ~

(a) We can grasp immediate causes and thus find them easiest to understand; this is why we like to think mechanistically about things which really are of a higher order.

(b) ... Thus mechanistic modes of explanation become the order of the day when we ignore problems which can only be explained dynamistically.

~ 5 ~

I will go so far as to assert ... that when an organism manifests itself we cannot grasp the unity and freedom of its formative impulse without the concept of metamorphosis.

~ 6 ~

Basic characteristic of an individual organism: to divide, to unite, to merge into the universal, to abide in the particular, to transform itself, to define itself, and, as living things tend to appear under a thousand conditions, to arise and vanish, to solidify and melt, to freeze and flow, to expand and contract. Since these effects occur together, any or all may occur at the same moment. Genesis and decay, creation and destruction, birth and death, joy and pain, all are interwoven with equal effect and weight; thus even the most isolated event always presents itself as an image and metaphor for the most universal.

~ 7 ~

Our ancestors admired the economy of Nature. She was thought to have a practical character, inclined to do much with small means where others produce little with great means. As mere mortals, we stand even more in admiration of the skill with which she is able to produce the widest variety of things while restricted to only a few basic principles.

To do this she uses the principle of life, with its inherent potential to work with the simplest phenomenon and diversify it by intensification into the most infinite and varied forms.

Whatever appears in the world must divide if it is to appear at all. What has been divided seeks itself again, can return to itself and reunite. This happens in a lower sense when it merely intermingles with its opposite, combines with it; here the phenomenon is nullified or at least neutralized. However, the union may occur in a higher sense if what has been divided is first intensified; then in the union of the intensified halves it will produce a third thing, something new, higher, unexpected.

~ *8* ~

Close observers of Nature, however diverse their points of view, will agree that everything of a phenomenal nature must suggest either an original duality capable of being merged in unity, or an original unity capable of becoming a duality. Separating what is united and uniting what is separate is the life of Nature. This is the eternal systole and diastole, the eternal *synkrisis* and *diakrisis,* the breathing in and out of the world in which we move and have our being.

~ *9* ~

In these notes entitled 'Analysis and Synthesis' (written in 1829, and first published posthumously in 1833), Goethe addresses the question of the place of analysis and synthesis in scientific work. As we have already seen, Goethe was no enemy of analytical research. He saw it as an important means of exposing false, and arriving at true, syntheses. However, he also argues that analysis alone is in vain unless it is undertaken within the overall context of establishing a synthetic understanding of the whole. In organic nature, the whole is always prior to the parts of which it is composed; while analytic work will give us a clear perception of the parts, be they ever so minute, the ultimate goal of scientific endeavour is to arrive at a perception of the underlying unity which binds them together.

In inorganic nature, this underlying unity is the law encapsulated in a theory or hypothesis. Goethe stresses the importance of testing the explanatory power of hypotheses by ranging against them as wide a variety of phenomena as possible. He targets Newton's method of using a single experimentum crucis *to prove a hypothesis as the antithesis of the correct scientific attitude. In this respect, Goethe is far more attuned to contemporary philosophy of science than Newton. But whereas contemporary philosophy of science tends towards the view that hypotheses are no more than conjectures, open to refutation by any succeeding analysis, for Goethe the synthetic activity of formulating general principles or laws works hand in hand with analytic activity. Syntheses are less conjectures than intuitions supported by, and arising out of, analytic work. The two kinds of activity are as inseparable as exhaling and inhaling.*

In this year's third lecture on the history of philosophy Mr. Victor Cousin* bestows high praise on the eighteenth century for its emphasis on the analytic method in science and for the care it took to avoid premature synthesis, that is, hypothesis. However, after giving almost unqualified approval to this approach, he notes that synthesis should not be excluded entirely since its use — albeit with caution — is sometimes necessary.

Consideration of these statements quickly led us to the thought that this is an area where the nineteenth century needs to do more: the friends and followers of science must note that we have failed to test, develop, and clarify false syntheses, that is, hypotheses handed down to us from the past. We have failed to restore to the human spirit its ancient right to come face to face with Nature.

Here we will cite two of these false syntheses by name: the decomposition of light and the polarization of light. Although often repeated by men of science, these two empty phrases say nothing to the thoughtful observer.

It is not enough that we apply the analytic approach to the observation of Nature; that is, that we refine as many details as possible out of a given object and thereby familiarize ourselves

* Victor Cousin, French philosopher (1792-1867).

with it. We should go on to apply the same analysis to existing syntheses so that we may discover whether a valid method has been applied in creating them.

Thus we have subjected Newton's approach to intensive analysis. He made the mistake of using a single phenomenon, and an over-refined one at that, as the foundation for a hypothesis supposed to explain the most varied and far-reaching events in Nature.

To develop our theory of colour we used the analytic approach; in so far as possible we presented every known phenomenon in a certain sequence so that we could determine the degree to which all might be governed by a general principle. It is our hope that this will help point the way for the nineteenth century as it carries out the duty we mentioned above.

We used a like approach in presenting the various phenomena created in double reflection. We bequeath both these efforts to some distant future in the knowledge that we have redirected our experiments to Nature and thus truly set them free.

* * *

Let us proceed to another more general observation. A century has taken the wrong road if it applies itself exclusively to analysis while exhibiting an apparent fear of synthesis: the sciences come to life only when the two exist side by side like exhaling and inhaling.

A false hypothesis is better than none at all, for the mere fact that it is false does no harm. But when such a hypothesis establishes itself, when it finds general acceptance and becomes something like a creed open to neither doubt nor test, it is an evil under which centuries to come will suffer.

Here Newton's theory may serve as an example. Objection to its shortcomings arose during Newton's own lifetime, yet these objections were smothered under the weight of his great accomplishments in other areas and his standing in social and learned circles. But the French are most to blame in disseminating and rigidifying this theory. It will be their task in the nineteenth century to rectify

their error by encouraging a fresh analysis of that tangled and
ossified hypothesis.

* * *

An important point is apparently overlooked when analysis is used
alone: every analysis presupposes a synthesis. A pile of sand
cannot be analysed, but if the pile contains grains of different
materials (sand and gold for instance), an analysis might be made
by washing it: then the light grains will wash away and the heavy
ones remain.

Thus modern chemistry depends largely on separating what
nature has united. We do away with nature's synthesis so that we
may learn about nature through its separate elements.

What higher synthesis is there than a living organism? Why
would we submit ourselves to the torment of anatomy, physiology,
and psychology if not to reach some concept of the whole, a concept which can always restore itself to wholeness no matter how it
is torn to pieces?

* * *

Therefore a great danger for the analytical thinker arises when he
applies his method where there is no underlying synthesis. In that
case his work will be a true labour of the Danaids,* and we can
find the saddest examples of this. For in essence he is simply
working to return to the synthesis. But if no synthesis underlies the
object of his attention he will labour in vain to discover it. All his
observations will only prove more and more an obstruction as their
number increases.

Thus the analytical thinker ought to begin by examining (or
rather, by noting) whether he is really working with a hidden
synthesis or only an aggregation, a juxtaposition, a composite, or
something of the sort. The areas of knowledge which have ceased

---

* The mythological daughters of Danaus who were condemned to pour water into
  a vessel with a sievelike bottom.

to develop raise such doubts. It might be possible to make some useful observations of this sort about the fields of geology and meteorology.

~ *10* ~

... separating and co-ordinating are two inseparable acts of life. Perhaps it is better to say that, whether we wish it or not, it is unavoidable for us to proceed from the whole to the parts and from the parts to the whole. And the more vitally these two functions of the mind are conjoined, like breathing in and out, the better it will be for science and its friends.

~ *11* ~

To rescue myself, I regard all phenomena as independent of one another and seek to isolate them at whatever cost; then I regard them as correlates, and they connect up in a decisive vital whole. I use this method above all in relation to nature; but also in relation to the latest events of world history that happen all around us ...

~ *12* ~

In this and the next quotation, we are introduced to the notions of polarity and intensification, which were fundamental to Goethe's thinking on morphology.

In Kant's scientific writings I had grasped the idea that attraction and repulsion are essential constituents of matter and that neither can be divorced from the other in the concept of matter. This led me to the recognition of polarity as a basic feature of all creation,

a principle permeating and animating the infinite range of phenomena.

~ *13* ~

The missing capstone* is the perception of the two great driving
forces in all Nature: the concepts of polarity and intensification, the
former a property of matter in so far as we think of it as material,
the latter in so far as we think of it as spiritual. Polarity is a state
of constant attraction and repulsion, while intensification is a state
of ever-striving ascent. Since, however, matter can never exist and
act without spirit, nor spirit without matter, matter is also capable
of undergoing intensification, and spirit cannot be denied its
attraction and repulsion. Similarly the capacity to think is given
only to someone who has made sufficient divisions to bring about
a union, and who has united sufficiently to seek further divisions.

~ *14* ~

*In the following classic statement, Goethe succinctly relates the
degree of perfection of an organism to the relationship of parts to
the whole.*

If we divide an organism into its anatomical parts and then in turn
divide these parts into their components, we finally come to such
beginnings as have been labelled 'similar parts.' We are not speaking of these here; we are rather pointing out a higher law of the
organism, which we shall explain as follows: Each living creature
is a complex, not a unit; even when it appears to be an individual,
it nevertheless remains an aggregation of living and independent

* Goethe is here commenting on the aphoristic essay 'Nature' by Georg Christoph
  Tobler which, because it appeared anonymously in Goethe's *Tiefurter Journal* (in
  the winter of 1782–83) has often wrongly been attributed to Goethe himself.

parts, identical in idea and disposition, but in outward appearance identical or similar, unlike or dissimilar. These organisms are partly united by origin; partly they discover each other and unite. They separate and seek each other out again, thus bringing about endless production in all ways and in all directions. The more imperfect a creature is, the more do these parts appear identical or similar to each other and the more do they resemble the whole. The more the creature is perfected, the more dissimilar its parts become. In the first case, the whole is more or less identical to the parts; in the second case, the whole is dissimilar to the parts. The more the parts resemble each other, the less they are subordinated to each other. Subordination of the parts betokens a more perfected creature.

~ *15* ~

In my opinion, the chief concept underlying all observation of life — one from which we must not deviate — is that a creature is self-sufficient, that its parts are inevitably interrelated, and that nothing mechanical, as it were, is built up or produced from without, although it is true that the parts affect their environment and are in turn affected by it.

~ *16* ~

Nothing is more consonant with Nature than that she puts into operation in the smallest detail that which she intends as a whole.

~ *17* ~

If you would seek comfort in the whole, you must learn to discover the whole in the smallest part.

~ 18 ~

Nothing happens in living Nature that does not bear some relation
to the whole ... The question is: how can we find the connection
between these phenomena, these events?

~ 19 ~

Above all we must remember that nothing that exists or comes into
being, lasts or passes, can be thought of as entirely isolated, entirely unadulterated. One thing is always permeated, accompanied,
covered, or enveloped by another; it produces effects and endures
them. And when so many things work through one another, where
are we to find the insight to discover what governs and what
serves, what leads the way and what follows? This creates great
difficulty in any theoretical statement: here lies the danger of
confusion between cause and effect, illness and symptom, deed and
character. Serious observers have no choice but to choose some
midpoint and then see how they can deal with what is left on the
periphery.

~ 20 ~

Whatever Nature undertakes, she can only accomplish it in a
sequence. She never makes a leap. For example she could not
produce a horse if it were not preceded by all the other animals on
which she ascends to the horse's structure as if on the rungs of a
ladder. Thus every one thing exists for the sake of all things and
all for the sake of one; for the one is of course the all as well.
Nature, despite her seeming diversity, is always a unity, a whole;
and thus, when she manifests herself in any part of that whole, the
rest must serve as a basis for that particular manifestation, and the
latter must have a relationship to the rest of the system.

~ *21* ~

The essay from which this extract is taken was probably dictated by Goethe in the early 1790's, but was never revised by Goethe himself. Found amongst the voluminous papers left at his death, 'Towards a General Comparative Theory' was first published posthumously in the Weimar edition of Goethes Werke.

The main argument of the essay is against the popular eighteenth century idea that God created the world for the benefit of human beings, and that therefore the task of science is to understand nature in terms of how it serves human purposes.

In this extract, Goethe digresses from his main argument in order to discuss one of his central concerns: the interaction between external environmental influences and factors intrinsic to an organism in determining the characteristic structure and form that a creature acquires. No creature develops in isolation from the environment in which it exists, and yet — in contrast to Darwinism — Goethe assumes that there is nevertheless an organizational principle within each creature. As we have seen in previous sections, this organizational principle should not be understood as something that can be isolated by physico-chemical analysis. It is more akin to a spiritual archetype than a genetic code, and Goethe would have no more embraced neo-Darwinism than he would have Darwinism. In the following extract, however, Goethe seeks to show that external, environmental factors do have an important role to play in determining forms, because all creatures exist in dynamic interrelationship with their environment.

In relating all things to themselves, human beings are forced to lend these things an inner purpose which is manifested externally, and all the more so because nothing alive can be imagined as existing without a complete structure. Since this complete structure develops inwardly in a fully specialized and specific way, it needs an external environment which is just as specialized. It can only exist in the outer world under certain conditions and in certain contexts.

Thus we find the most varied forms of animal life stirring on the earth, in the water, and in the air. The common view is that these creatures have received their appendages for the purpose of making various movements and thereby supporting their particular form of existence. But will we not show more regard for the primal force of Nature, for the wisdom of the intelligent being usually presumed to underlie it, if we suppose that even its power is limited, and realize that its forms are created by something working from without as well as from within? The statement 'The fish exists for the water' seems to me to say far less than 'The fish exists in the water and by means of the water.' The latter expresses more clearly what is obscured in the former; that is, the existence of a creature we call 'fish' is only possible under the conditions of an element we call 'water,' so that the creature not only exists in that element, but may also evolve there.

The same principle holds true of all other creatures. An initial and very general observation on the outer effect of what works from within and the inner effect of what works from without would therefore be as follows: the structure in its final form is, as it were, the inner nucleus moulded in various ways by the characteristics of the outer element. It is precisely thus that the animal retains its viability in the outer world: it is shaped from without as well as from within. And this is all the more natural because the outer element can shape the external form more easily than the internal form. We can see this most clearly in the various species of seal, where the exterior has grown quite fishlike even though the skeleton still retains all the features of a quadruped.

We show disrespect neither for the primal force of Nature nor for the wisdom and power of a creator if we assume that the former acts indirectly, and that the latter acted indirectly at the beginning of all things. Is it not fitting that this great force should bring forth simple things in a simple way and complex things in a complex way? Do we disparage its power if we say it could not have brought forth fish without water, birds without air, other animals without earth, that this is just as inconceivable as the continued existence of these creatures without the conditions provided by each element? Will we not attain a more satisfactory insight into the mysterious architecture of the formative process, now widely recognized to be built on a single pattern, by examining and comprehending this single pattern more fully and then looking into the following question: how does a surrounding element, with its various specific characteristics, affect the general form we have been studying? How does the form, both determined and a determinant, assert itself against these elements? What manner of hard parts, soft parts, interior parts, and exterior parts are created in the form by this effect? And, as indicated before, what is wrought by the elements through all their diversity of height and depth, region and climate?

Much research has already been done on these points. This needs only to be brought together and applied, but in accordance with the method described above.

How admirable that Nature must use the same means to produce a creature as it does to sustain it! We progress on our path as follows: first we viewed the unstructured, unlimited element as a vehicle for the unstructured being, and now we will raise our observation to a higher level to consider the structured world itself as an interrelationship of many elements. We will see the entire plant world, for example, as a vast sea which is as necessary to the existence of individual insects as the oceans and rivers are to the existence of individual fish, and we will observe that an enormous number of living creatures are born and nourished in this ocean of plants. Ultimately we will see the whole world of animals as a great element in which one species is created, or at least sustained, by and through another. We will no longer think of connections and relationships in terms of purpose and intention. This is the only road to progress in understanding how Nature expresses itself from all quarters and in all directions as it goes about its work of creation. As we find through experience, and as the advance of science has shown, the most concrete and far-reaching benefits for humanity come from an intense and selfless effort which neither demands its reward at week's end like a labourer, nor lies under any obligation to produce some useful result for mankind after a year, a decade, or even a century.

4. Quality and Quantity: Two Poles of Material Existence

Goethe's morphological approach to nature is qualitative rather than quantitative. While he has great respect for mathematics, Goethe believes that it cannot give us a complete account of reality, since its province is restricted to the measurable, and the qualitative aspect of nature is not susceptible of measurement. Since mathematics is capable of dealing only with that aspect of nature which can be quantified, it is necessarily marginal to morphology, as conceived by Goethe. A science of nature based on mathematical methods alone will result in a partial view of nature that would need to be complemented by a qualitative science such as Goethe advocates.

It would be wrong, however, to attribute to Goethe a hostility towards mathematics as such. What he opposes is the insinuation of mathematics into the very fabric of scientific knowledge, with the consequence not only of a partial view of nature, but also of a knowledge that is separated off from the whole human being. In itself, mathematics is, of course, indispensable for grasping all that is quantifiable in nature. However, the qualitative aspect of nature has primacy over the quantitative, since it (that is. the qualitative) is what is given to our immediate experience. As human beings in relationship to the world, our actual experience is of qualities, not of numbers or mathematical formulas. The latter rest upon a previous abstraction of quantities from experienced qualities, and it is for this reason that Goethe sees the non-mathematical grasp of the qualitative in nature as having priority over mathematical analysis.

~ *1* ~

The mathematician relies on the element of quantity, on all that is defined by number and size, and thus to some degree on the universe in its external form. But if we set out to apply the full measure of mind and all its powers to this universe, we will realize that quantity and quality must be viewed as two poles of material existence. This is why the mathematician refines his language of formula so highly; as far as possible he wants to incorporate the incalculable world into the realm of measure and number. Everything will then seem graspable, comprehensible, and mechanical, and he may be accused of an underlying atheism for supposedly he has included the most incalculable element of all (which we call God), and thus has eliminated its special, overriding presence.

~ *2* ~

A strict separation must be maintained between the physical sciences and mathematics. The physical sciences must remain quite independent; they must use all their powers of love, respect, and reverence to find their way into Nature and the sacred life of Nature irrespective of what mathematics does. The latter, on the other hand, must declare itself independent of all externalities, take its own path of intellect, and develop in a purer way than it now does in working with the physical world to gain something from it or impose something on it.

~ *3* ~

An important task: to banish mathematical-philosophical theories from those areas of physical science where they impede rather than advance knowledge, those areas where a one-sided development in modern scientific education has made such perverse use of them.

~ *4* ~

It does not at all follow that the hunter, who has killed the game, must also be the cook who prepares it. Perchance a cook can go along on the hunt and be a good shot; but one would commit a bad fallacy if one were to maintain that in order to be a good shot, one must be a cook. Mathematicians appear to me to be in the same case when they maintain that one can see nothing, find nothing, in the physical realm without being a mathematician, since they might be well satisfied if one were to bring to them in the kitchen the kind of thing they can lard with formulas and prepare in their own way.

~ *5* ~

Number and measurement in all their baldness destroy form and banish the spirit of living contemplation.

~ *6* ~

I receive mathematics as the most sublime and useful science, so long as they are applied in their proper place; but I cannot commend the misuse of them in matters which do not belong to their sphere, and in which, noble science as they are, they seem to be mere nonsense. As if, forsooth! things only exist when they can be mathematically demonstrated. It would be foolish for a man not to believe in his mistress' love because she could not prove it to him mathematically. She can mathematically prove her dowry, but not her love. The mathematicians did not find out the metamorphosis of plants. I have achieved this discovery without the aid of mathematics, and the mathematicians were forced to put up with it. To understand the phenomena of colour, nothing is required but unbiased observation and a sound head; but these are scarcer than folks imagine.

~ *7* ~

It is a false notion that a phrase of a mathematical formula can ever take the place of, or set aside, a phenomenon.

~ *8* ~

The process of measuring is a coarse one, and extremely imperfect when applied to a living object. A living thing cannot be measured by something external to itself; if it must be measured, it must provide its own gauge. This gauge, however, is highly spiritual, and cannot be formed through the senses.

~ *9* ~

It is generally agreed that, while mathematics in itself can be treated purely and give assured results, it runs into constant danger when it gets into the terrain of sense-experience.

~ *10* ~

From the mathematician, we must learn the meticulous care required to connect things in unbroken succession, or rather, to derive things step by step. Even where we do not venture to apply mathematics, we must always work as though we had to satisfy the strictest of geometricians.

~ *11* ~

What except for its exactitude is exact about mathematics? And
this exactitude — does it not flow from an inner feeling for the
truth?

~ *12* ~

Mathematics cannot eliminate prejudice, prevent wilfulness, or
resolve partisan differences. It has no power over anything in the
moral realm.

~ *13* ~

(a) Like dialectics, mathematics is an organ for a higher kind of
inner sense; in practice it is an art like rhetoric. Both value nothing
but form — the content is unimportant. It does not matter whether
mathematics counts pennies or guineas, whether rhetoric defends
what is true or what is false.

(b) Here, however, the character of the person doing these things,
practising these arts, is most important. An effective advocate with
a just cause, an able mathematician before the starry heavens —
both seem equally godlike.

~ *14* ~

Mathematicians are perfect only to the degree that they are perfect
human beings, to the degree that they can experience the beauty in
what is true. Only then will their work be complete, transparent,
comprehensive, pure, clear, graceful — even elegant.

~ *15* ~

I have heard myself criticized as if I were an opponent, an enemy, of mathematics in general, which in fact no one can value more highly than I.

5. Delicate Empiricism

In this and the next two chapters, Goethe's leading thoughts on scientific method are presented. His method is characterized by a 'soft' approach to nature, in which the scientist works from an attitude of receptive 'listening,' rather than an overactive conjecturing combined with attempts to either prove or disprove the conjectures. Goethe seeks instead to attune to what is experienced, refraining as far as possible from trying to fit the experience into any preconceived ideas or theories. His method rests on the premiss that ideas and theories will emerge as an implicit part of a deepened experience of the phenomena under investigation.

In this respect, Goethe's method is neither inductivist (that is, proceeding from particular sense-experiences to universal explanatory principles or laws), nor is it hypothetico-deductive (creating provisional hypotheses from which various phenomena may be deduced, but which always remain open to refutation by subsequent, unpredicted phenomena). Broadly speaking, inductivism was the official methodology of mainstream science through most of the nineteenth century, whereas hypothetico-deductivism has dominated twentieth century philosophy of science. The problem with inductivism is that it makes an artificial division between the act of observation and the creation of an explanatory principle. In reality, it is impossible to make 'pure' observations which have no thought-content whatsoever. All empirical observations assume a certain theoretical stance on the part of the observer, even though this may be relatively unconscious.

But long before the hypothetico-deductivists attacked the inductivists on the grounds that all observation is 'theory-laden,' Goethe himself had insisted that 'the realm of fact is already theory' (see Chapter 6, 6). This, however, did not lead Goethe to embrace the hypothetico-deductive method. The essential feature of this method is that it denies any cognitive value to the formation of hypotheses in order to throw emphasis on the value of attempts to refute a given hypothesis by appealing to phenomena which it cannot adequately account for. This is why every scientific theory (according to the hypothetico-deductivists) remains merely provisional, for the actual formation of hypotheses exists apart from the process of refutation. How a hypothesis is formed does not give to it any intrinsic merit: its merit lies only in whether or not it can withstand repeated attempts at refutation.

What links inductivism and hypothetico-deductivism is the assumption that the human mind necessarily operates in a separate sphere from nature, formulating its ideas and theories from a position of relative detachment from phenomena. Goethe's methodology, by contrast, is based on the assumption that the human mind has the capacity to enter into the essence of phenomena, and lay hold of principles which, once grasped, alter one's perception of — and relationship to — these phenomena.

Goethe aptly described his method as a 'delicate empiricism' (Zarte-empirie). It is painstakingly attentive to the phenomena, and requires from the scientist the discipline of always taking the lead from the phenomena, allowing them to speak, and silencing the scientist's own urge to rush into premature explanatory hypotheses. The Goethean scientist seeks to participate in the objects investigated to such a degree that the mind makes itself one with the object, thereby overcoming the sense of separateness that characterizes our normal experience of ourselves in relation to the world. It is in this respect that Goethe's methodology differs radically from both inductivism and hypothetico-deductivism. For the essential plasticity of the act of observation means that it can be shaped and guided by the mind's attentiveness to what the phenomena are really saying. Thus it is possible to arrive at a theory that is at the same time a genuine insight into the phenomena investigated. The theory is not something apart from the experience of the phenomena, but informs and enriches the experience, thus furthering an enhanced sensitivity towards the natural world.

~ *1* ~

There is a delicate empiricism which makes itself utterly identical
with the object, thereby becoming true theory. But this enhancement of our mental powers belongs to a highly evolved age.

~ *2* ~

Nature will reveal nothing under torture; its frank answer to an
honest question is 'Yes! Yes! — No! No!' More than this comes
of evil.

~ *3* ~

In the sciences, we find ... innumerable attempts to systematize, to
schematize. But our full attention must be focused on the task of
listening to Nature to overhear the secret of her process, so that we
neither frighten her off with coercive imperatives, nor allow her
whims to divert us from our goal.

~ *4* ~

The manifestation of a phenomenon is not independent of the
observer — it is caught up and entangled in his individuality.

~ *5* ~

In observing Nature on a scale large or small, I have always asked:
Who speaks here, the object or you? I also take this approach in
regard to my predecessors and colleagues.

~ *6* ~

... Thus when making observations it is best to be fully conscious of objects, and when thinking to be fully aware of ourselves.

~ *7* ~

The present age has a bad habit of being abstruse in the sciences. We remove ourselves from common sense without opening up a higher one; we become transcendent, fantastic, fearful of intuitive perception in the real world, and when we wish to enter the practical realm, or need to, we suddenly turn atomistic and mechanical.

~ *8* ~

My whole method relies on derivation. I persist until I have discovered a pregnant point from which several things may be derived, or rather — since I am careful in my work and observations — one which yields several things, offering them up of its own accord. If some phenomenon appears in my research, and I can find no source for it, I let it stand as a problem. This approach has proven quite advantageous over the years. The origin and context of some problem might be impossible to discover; I might have to let it lie for a long time; but at some moment, years later, enlightenment comes in the most wonderful way.

~ 9 ~

Written in 1792 (though not published until 1823), 'The Experiment as Mediator between Subject and Object' is one of the fullest and most important of Goethe's statements on scientific method. It shows his dedication to the goal of selfless observation of nature, in which the tendencies of the ego to have its own predilections and preconceptions confirmed are constantly checked.

The thrust of the essay is directed against Newton's conception of the role of the single experimentum crucis *(crucial experiment), designed to test the truth of a scientific hypothesis. In the ex-*perimentum crucis *the experimenter deliberately creates conditions which isolate the critical factors on which the hypothesis depends, in order to test its veracity. Thus from a single experiment, Newton claimed to prove that colour is produced by the different refrangibility of rays of light. According to Goethe, Newton's* experimentum crucis *simply compelled nature to confirm his own preconceived ideas, by producing, under constraint, the very phenomena Newton was wanting to find.*

For Goethe, the single experiment has little value in isolation from other experiments, and should never be taken as proof of an hypothesis: it simply creates the conditions under which a certain set of phenomena arise, which should then be connected with a host of other phenomena produced under other conditions. The value of experimentation thus lies in the experimental series *rather than in any one isolated experiment. For Goethe, the question is always: how are we to find the connection between diverse phenomena? The answer is: by conducting a series of experiments in which these phenomena, viewed from a multitude of different aspects and manifesting under varying conditions, are able to* reveal *an underlying connectedness. The inner unity of the series of experiments can then be experienced: as it were, a single experience in many diverse aspects. Once again, the aim is not so much to devise a theory or explanatory system, but rather to attain a 'higher' experience of the unity underlying a diversity of perceptions.*

THE EXPERIMENT AS MEDIATOR BETWEEN SUBJECT AND OBJECT

(1) As we become aware of objects in our environment we will relate them to ourselves, and rightly so since our fate hinges on whether these objects please or displease us, attract or repel us, help or harm us. This natural way of seeing and judging things seems as easy as it is essential, although it can lead to a thousand errors — often the source of humiliation and bitterness in our life.

(2) A far more difficult task arises when a person's thirst for knowledge kindles in them a desire to view Nature's objects in their own right and in relation to one another. On the other hand we lose the yardstick which came to our aid when we looked at things from the human standpoint; that is, in relation to ourselves. This yardstick of pleasure and displeasure, attraction and repulsion, help and harm, we must now renounce absolutely; as a neutral, seemingly godlike being we must seek out and examine what it is, not what pleases. Thus true botanists must remain unmoved by beauty or utility in a plant; they must explore its formation, its relation to other plants. Like the sun which draws forth every plant and shines on all, they must look upon each plant with the same quiet gaze; they must find the measure for what they learn, the data for judgment, not in themselves but in the sphere of what they observe.

(3) The history of science teaches us how difficult this renunciation is for human beings.. The second part of our short essay will discuss how we thus arrive (and must arrive) at hypotheses, theories, systems, any of the modes of perception which help in our effort to grasp the infinite; the first part of the essay will deal with how we set about recognizing the forces of Nature. Recently I have been studying the history of physics and this point arose frequently — hence the present brief discourse, an attempt to outline in general how the study of Nature has been helped or hindered by the work of able scientists.

(4) We may look at an object in its own context and the context of other objects, while refraining from any immediate response of desire or dislike. The calm exercise of our powers of attention will quickly lead us to a rather clear concept of the object, its parts, and

its relationships; the more we pursue this study, discovering further relations among things, the more we will exercise our innate gift of observation. Those who understand how to apply this knowledge to their own affairs in a practical way are rightly deemed clever. It is not hard for any well-organized person, moderate by nature or force of circumstance, to be clever, for life corrects us at every step. But if we are called upon to apply this keen power of judgment to exploring the hidden relationships in nature, if we are to find our own way in a world where we are seemingly alone, if we are to avoid hasty conclusions and keep a steady eye on the goal while noting every helpful or harmful circumstance along the way, if we must be our own sharpest critics where no one else can test our work with ease, if we must question ourselves continually even when most enthusiastic — it is easy to see how harsh these demands are and how little hope there is of seeing them fully satisfied in ourselves or others. Yet these difficulties, this hypothetical impossibility, must not deter us from doing what we can. At any rate, our best approach is to recall how able people have advanced the sciences, and to be candid about the false paths down which they have strayed, only to be followed by numerous disciples, often for centuries, until later empirical evidence could bring researchers back to the right road.

(5) It is undeniable that in the science now under discussion, as in every human enterprise, empirical evidence carries (and should carry) the greatest weight. Neither can we deny the high and seemingly creative independent power found in the inner faculties through which the evidence is grasped, collected, ordered, and developed. But how to gather and use empirical evidence, how to develop and apply our powers — this is not so generally recognized or appreciated.

(6) We might well be surprised how many people are capable of sharp observation in the strictest sense of the word. When we draw their attention to objects, we will discover that such people enjoy making observations, and show great skill at it. Since taking up my study of light and colour I have often had opportunity to appreciate this. Now and then I discuss my current interests with people unacquainted with the subject; once their attention is awakened they

frequently make quick note of phenomena I was unaware of or had neglected to observe. Thus they may be able to correct ideas developed in haste, and even produce a breakthrough by transcending the inhibitions in which exacting research often traps us.

(7) Thus what applies in so many other human enterprises is also true here: the interest of many focused on a single point can produce excellent results. Here it becomes obvious that researchers will meet their downfall if they have any feeling of envy which seeks to deprive others of the discoverer's laurels, any overwhelming desire to deal alone and arbitrarily with a discovery.

(8) I have always found the cooperative method of working satisfactory, and I intend to continue with it. I am aware of the debts I have incurred along the way, and it will give me great pleasure later to acknowledge these publicly.

(9) If people's natural talent for observation can be of such help to us, how much more effective must it be when trained observers work hand in hand. In and of itself, a science is sufficient to support the work of many people, although no one person can carry an entire science. We may note that knowledge, like contained but living water, rises gradually to a certain level, and that the greatest discoveries are made not so much by individuals as by the age; important advances are often made by two or more skilled thinkers at the same time. We have already found that we owe much to the community and our friends; now we discover our debt to the world and the age we live in. In neither case can we appreciate fully enough our need for communication, assistance, admonition, and contradiction to hold us to the right path and help us along it.

(10) Thus in scientific matters we must do the reverse of what is done in art. Artists should never present a work to the public before it is finished because it is difficult for others to advise or help them with its production. Once it is finished, however, they must consider criticism or praise, take it to heart, make it a part of their own experience, and thereby develop and prepare themselves for new works. In science, on the other hand, it is useful to publish every bit of empirical evidence, even every conjecture; indeed, no scientific edifice should be built until the plan and materials of its structure have been widely known, judged and sifted.

(11) I will now turn to a point deserving of attention; namely, the method which enables us to work most effectively and surely.

(12) When we intentionally reproduce empirical evidence found by earlier researchers, contemporaries, or ourselves, when we re-create natural or artificial phenomena, we speak of this as an experiment.

(13) The main value of an experiment lies in the fact that, simple or compound, it can be reproduced at any time given the requisite preparations, apparatus, and skill. After assembling the necessary materials we may perform the experiment as often as we wish. We will rightly marvel at human ingenuity when we consider even briefly the variety of arrangements and instruments invented for this purpose. In fact, we can note that such instruments are still being invented daily.

(14) As worthwhile as each individual experiment may be, it receives its real value only when united or combined with other experiments. However, to unite or combine just two somewhat similar experiments calls for more rigor and care than even the sharpest observers usually expect of themselves. Two phenomena may be related, but not nearly so closely as we think. Although one experiment seems to follow from another, an extensive series of experiments might be required to put the two into an order actually conforming to nature.

(15) Thus we can never be too careful in our efforts to avoid drawing hasty conclusions from experiments or using them directly as proof to bear out some theory. For here at this pass, this transition from empirical evidence to judgment, cognition to application, all the inner enemies of humanity lie in wait: imagination, which sweeps us away on its wings before we know our feet have left the ground; impatience; haste; self-satisfaction; rigidity; formalistic thought; prejudice; ease; frivolity; fickleness — this whole throng and its retinue. Here they lie in ambush and surprise not only the active observer but also the contemplative one who appears safe from all passion.

(16) I will present a paradox of sorts as a way of alerting the reader to this danger, far greater and closer at hand than we might think. I would venture to say that we cannot prove anything by one

experiment or even several experiments together, that nothing is more dangerous than the desire to prove some thesis directly through experiments, that the greatest errors have arisen just where the dangers and shortcomings in this method have been overlooked. I will explain this assertion more clearly lest I merely seem intent on raising a host of doubts. Every piece of empirical evidence we find, every experiment in which this evidence is repeated, really represent just one part of what we know. Through frequent repetition we attain certainty about this isolated piece of knowledge. We may be aware of two pieces of empirical evidence in the same area; although closely related, they may seem even more so, for we will tend to view them as more connected than they really are. This is an inherent part of our nature; the history of human understanding offers thousands of examples of this, and I myself make this error almost daily.

(17) This mistake is associated with another which often lies at its root. Human beings take more pleasure in the idea than in the thing; or rather, we take pleasure in a thing only in so far as we have an idea of it. The thing must fit our character, and no matter how exalted our way of thinking, no matter how refined, it often remains just a way of thinking, an attempt to bring several objects into an intelligible relationship which, strictly speaking, they do not have. Thus the tendency to hypotheses, theories, terminologies, and systems, a tendency altogether understandable since it springs by necessity from the organization of our being.

(18) Every piece of empirical evidence, every experiment, must be viewed as isolated, yet the human faculty of thought forcibly strives to unite all external objects known to it. It is easy to see the risk we run when we try to connect a single bit of evidence with an idea already formed, or use individual experiments to prove some relationship not fully perceptible to the senses but expressed through the creative power of the mind.

(19) Such efforts generally give rise to theories and systems which are a tribute to their author's intelligence. But with undue applause or protracted support they soon begin to hinder and harm the very progress of the human mind they had earlier assisted.

(20) We often find that the more limited the data, the more artful

a gifted thinker will become. As though to assert his sovereignty he chooses a few agreeable favourites from the limited number of facts and skilfully marshals the rest so they never contradict him directly.* Finally he is able to confuse, entangle, or push aside the opposing facts and reduce the whole to something more like the court of a despot than a freely constituted republic.

(21) So deserving a man will not lack admirers and disciples who study this fabric of thought historically, praise it, and seek to think as much like their master as possible. Often such a doctrine becomes so widespread that anyone bold enough to doubt it would be considered brash and impertinent. Only in later centuries would anyone venture to approach such a holy relic, apply common sense to the subject, and — taking a lighter view — apply to the founder of the sect what a wag once said of a renowned scientist: 'He would have been a great man if only he hadn't invented so much.'

(22) It is not enough to note this danger and warn against it. We need to declare our own views by showing how we ourselves would hope to avoid this pitfall, or by telling what we know of how some predecessor avoided it.

(23) Earlier I stated my belief that the direct use of an experiment to prove some hypothesis is detrimental; this implies that I consider its indirect use beneficial. Here we have a pivotal point, one requiring clarification.

(24) Nothing happens in living Nature that does not bear some relation to the whole. The empirical evidence may seem quite isolated, we may view our experiments as mere isolated facts, but this is not to say that they are, in fact, isolated. The question is: how can we find the connection between these phenomena, these events?

(25) Earlier we found those thinkers most prone to error who seek to incorporate an isolated fact directly into their thinking and judgment. By contrast, we will find that the greatest accomplishments come from those who never tire in exploring and working out every possible aspect and modification of every bit of empirical evidence, every experiment.

* Goethe has Newton in mind.

(26) It would require a second essay to describe how our intellect can help us with this task; here we will merely indicate the following. All things in Nature, especially the commoner forces and elements, work incessantly upon one another; we can say that each phenomenon is connected with countless others just as we can say that a point of light floating in space sends its rays in all directions. Thus when we have done an experiment of this type, found this or that piece of empirical evidence, we can never be careful enough in studying what lies next to it or derives directly from it. This investigation should concern us more than the discovery of what is related to it. To follow every single experiment through its variations is the real task of scientific researchers. Their duty is precisely the opposite of what we expect from authors who write to entertain. The latter will bore their readers if they do not leave something to the imagination, while the former must always work as if they wished to leave nothing for their successors to do. Of course, the disproportion between our intellect and the nature of things will soon remind us that no one has gifts enough to exhaust the study of any subject.

(27) In the first two parts of my *Contributions to Optics,* I sought to set up a series of contiguous experiments derived from one another in this way. Studied thoroughly and understood as a whole, these experiments could even be thought of as representing a single experiment, a single piece of empirical evidence explored in its most manifold variations.

(28) Such a piece of empirical evidence, composed of many others, is clearly of a higher sort. It shows the general formula, so to speak, that overarches an array of individual arithmetic sums. In my view, it is the task of the scientific researcher to work toward empirical evidence of this higher sort — and the example of the best workers in the field supports this view. From the mathematician we must learn the meticulous care required to connect things in unbroken succession, or rather, to derive things step by step. Even where we do not venture to apply mathematics we must always work as though we had to satisfy the strictest of geometricians.

(29) In the mathematical method we find an approach which by

its deliberate and pure nature instantly exposes every leap in an assertion. Actually, its proofs merely state in a detailed way that what is presented as connected was already there in each of the parts and as a consecutive whole, that it has been reviewed in its entirety and found to be correct and irrefutable under all circumstances. Thus its demonstrations are always more exposition, recapitulation, than argument. Having made this distinction, I may now return to something mentioned earlier.

(30) We can see the great difference between a mathematical demonstration which traces the basic elements through their many points of connection, and the proof offered in the arguments of a clever speaker. Although arguments may deal with utterly separate matters, wit and imagination can group them around a single point to create a surprising semblance of right and wrong, true and false. It is likewise possible to support a hypothesis or theory by arranging individual experiments like arguments and offering proofs which bedazzle us to some degree.

(31) But those who wish to be honest with themselves and others will try by careful development of individual experiments to evolve empirical evidence of the higher sort. These pieces of evidence may be expressed in concise axioms and set side by side, and as more of them emerge they may be ordered and related. Like mathematical axioms they will remain unshakable either singly or as a whole. Anyone may examine and test the elements, the many individual experiments, which constitute this higher sort of evidence; it will be easy to judge whether we can express these many components in a general axiom, for nothing here is arbitrary.

(32) The other method which tries to prove assertions by using isolated experiments like arguments often reaches its conclusions furtively or leaves them completely in doubt. Once sequential evidence of the higher sort is assembled, however, our intellect, imagination and wit can work upon it as they will; no harm will be done, and, indeed, a useful purpose will be served. We cannot exercise enough care, diligence, strictness, even pedantry, in collecting basic empirical evidence; here we labour for the world and the future. But these materials must be ordered and shown in sequence, not arranged in some hypothetical way nor made to serve

the dictates of some system. Everyone will then be free to connect them in his own way, to form them into a whole which brings some measure of delight and comfort to the human mind. This approach keeps separate what must be kept separate; it enables us to increase the body of evidence much more quickly and cleanly than the method which forces us to cast aside later experiments like bricks brought to a finished building.

(33) The views and examples of the best scientists give me reason to hope that this is the right path, and I trust my explanation will satisfy those of my friends who ask from time to time what I am really seeking to accomplish with my optical experiments. My intention is to collect all the empirical evidence in this area, do every experiment myself, and develop the experiments in their most manifold variations so that they become easy to reproduce and more accessible. I will then attempt to establish the axioms in which the empirical evidence of a higher nature can be expressed, and see if these can be subsumed under still higher principles. If imagination and wit sometimes run impatiently ahead on this path, the method itself will fix the bounds to which they must return.

April 28, 1792

~ 10 ~

... it is the observer's first duty to discover every condition under which a phenomenon may occur, and to aim at the completeness of the phenomena, since they actually form a series, or rather are forced to interpenetrate, so that they will present themselves to one's observation as an organization manifesting an inner life of its own.

~ *11* ~

No phenomenon can be explained by itself and of itself; only a number of them, when viewed collectively, and arranged methodically, end by yielding something that may pass for theory.

~ *12* ~

No hypothesis can lay claim to any value unless it assembles many phenomena under one concept.

~ *13* ~

If we hold fast to the regularity we will always find ourselves led back to it by our observations; anyone who fails to recognize the law will doubt the phenomenon, for, in the highest sense, every exception is included in the rule.

~ *14* ~

Hypotheses are lullabies with which the teacher soothes his pupils to sleep. The thoughtful and faithful observer grows increasingly conscious of his limitations, for he perceives that the more knowledge extends the more numerous are the problems that emerge.

~ *15* ~

Content without method leads to phantasy; method without content to empty sophistry; matter without form to unwieldy erudition; form without matter to hollow speculation.

~ *16* ~

People who behold a phenomenon will often extend their thinking beyond it; people who merely hear about the phenomenon will not be moved to think at all.

~ *17* ~

Throughout the history of scientific investigation we find observers leaping too quickly from phenomenon to theory; hence they fall short of the mark and become theoretical.

~ *18* ~

All hypotheses get in the way of the *anatheorismos* — the urge to look again, to contemplate the objects, the phenomena in question, from all angles.

~ *19* ~

Hypotheses are scaffoldings that one erects in advance of the building and that one takes down when the building is finished. The worker cannot do without them. But he must be careful not to mistake the scaffolding for the building.

~ 20 ~

Theory in and of itself is of no use, except in as far as it makes us believe in the interconnection of phenomena.

~ 21 ~

When one considers the problems of Aristotle, one is astonished at his gift of observation, and at all that the Greeks had an eye for; only they err in being over-hasty, for they go directly from the phenomenon to the explanation, thereby producing very inadequate theoretical conclusions. This, however, is a general mistake that is still made, even in our own day.

~ 22 ~

Written in 1807, shortly after the French occupation of Weimar in 1806, this short essay formed the introduction to the first volume of Natural Science in General; Morphology in Particular (1817). In it Goethe indicates how his own approach to nature steers a middle course between empiricism and idealism; or rather, seeks a dynamic combination of both these tendencies.

OUR UNDERTAKING IS DEFENDED

When people of lively intellect first respond to Nature's challenge to be understood, they feel irresistibly tempted to impose their will upon the natural objects they are studying. Before long, however, these natural objects close in upon us with such force as to make us realize that we in turn must now acknowledge their might and hold in respect the authority they exert over us. Hardly are we convinced of this reciprocal influence when we become aware of a twofold infinitude: in the natural objects, of the diversity of life and growth and of vitally interlocking relationships; in ourselves, of the possibility of endless development through always keeping our minds receptive and disciplining our minds in new forms of assimilation and procedure.

These circumstances give high pleasure and would insure one's happiness in life if obstacles from within and without did not block the beautiful path of perfection. The years, which at first gave freely, now begin to take their toll; we are satisfied, each in our own measure, with what has been accomplished, and we are quite content to rejoice over it in private, inasmuch as genuine, pure, and stimulating sympathy on the part of others is rare.

How few are inspired by what is perceptible to the intellect alone! The senses, the emotions, the soul exert far greater power over us, and rightly so, since we are destined for active life and not for meditation.

Alas, even in those devoted to understanding and knowledge we seldom find the interest we hope for! Pragmatists — taking note of the special case, observing and analysing in detail — are inclined to regard as an encumbrance those things that stem from and lead back again to an idea. In their way they feel at home in their labyrinth and take no interest in a thread that would guide them, not only more rapidly but all the way through. To such people, metal that is unminted and cannot be counted is a burdensome possession. On the other hand, people of wider horizons are all too inclined to disdain and to reduce to a deadening generality what possesses vitality only as a particular.

In this conflict we have long been engaged. Much has been accomplished in the process, much has been destroyed; and I should never be tempted to entrust my views on Nature to the seas of opinion, in so weak a craft, if we had not had occasion in recent hours of danger to feel all too keenly the value of papers on which we have felt impelled to set down a part of our being.*

Let, therefore, what in dauntless youth I had often dreamed of as a finished work, now go forth as a sketch, as a mere fragmentary collection, to function and serve for what it is.

---

* Goethe is referring to the looting of Weimar by the French in October 1806.

This much it was necessary for me to say in commending these early sketches — some individual parts of which, however, are more or less complete — to the good will of my contemporaries. A good many things that may still remain to be said, we had best introduce in the course of our undertaking.

Jena, 1807

6. The Contemplation of Nature Suggests Ideas

If the tendency to jump too quickly to hypotheses and theories is one danger the Goethean scientist seeks to avoid, this is not because Goethe viewed human thinking as necessarily or irrevocably subjective. Rather, he believed that nature herself works according to ideas, and the task of the scientist is to avoid placing too great an emphasis on hypothesizing precisely in order to keep the mind open to perceiving the ideas operative in nature. Goethe's view of our ideas as having — at least potentially — an objective basis in nature places him in a Platonic as opposed to a Kantian current of thinking, and hence at odds with much contemporary philosophy of science.

Knowledge, for Goethe, is not arrived at by imposing ideas on experiences, but by deepening experiences to the point at which their innate idea-content is made manifest. Our thoughts then arise out of the empirical experience as a further dimension of that experience, not simply as a more or less subjective and utilitarian ordering of experience.

The contemplation of nature will suggest to us ideas because a truly contemplative observation of nature will lead to an intuitive insight into that which works creatively and spiritually within natural phenomena. The present chapter is crucial to an understanding of the philosophical basis of Goethe's scientific method, and provides the foundations of his concept of the Archetypal Phenomenon discussed in the following chapter.

~ *1* ~

That the observation of Nature leads to thinking; that its abundance makes us resort to a variety of methods in order to manipulate it even to some degree — on this there seems to be general agreement. But only a limited few are equally aware of the fact that the contemplation of Nature suggests ideas to which we ascribe the same degree of certainty as to nature itself — a greater degree, in fact; and that we have a right to be guided by these ideas both in our search for data and in our attempts to arrange what we have found.

~ *2* ~

An extremely odd demand is often set forth but never met, even by those who make it: that is, that empirical data should be presented without any theoretical context, leaving the reader, the student, to his own devices in judging it. This demand seems odd because it is useless simply to look at something. Every act of looking turns into observation, every act of observation into reflection, every act of reflection into the making of associations; thus it is evident that we theorize every time we look carefully at the world. The ability to do this with clarity of mind, with self-knowledge, in a free way, and (if I may venture to put it so) with irony, is a skill we will need in order to avoid the pitfalls of abstraction and attain the results we desire, results which can find a living and practical application.

~ *3* ~

Two needs arise in us when we observe Nature: to gain complete knowledge of the phenomena themselves, and then to make them our own by reflection upon them. Completeness is a product of

order, order demands method, and method makes it easier to perceive the concept. When we are able to survey an object in every
detail, grasp it correctly, and reproduce it in our mind's eye, we
can say that we have an intuitive perception of it in the truest and
highest sense. We can say that it belongs to us, that we have
attained a certain mastery of it. And thus the particular always
leads us to the general, the general to the particular. The two combine their effects in every observation, in every discourse.

~ *4* ~

The particular eternally underlies the general *(Allgemeine);* the
general has eternally to adapt itself to the particular.

~ *5* ~

We must learn to see that what we have seen and recognized in the
most simple must also be supposed and believed in the complex.
For the simple conceals itself in the manifold.

~ *6* ~

The ultimate goal would be: to grasp that everything in the realm
of fact is already theory. The blue of the sky shows us the basic
law of chromatics. Let us not seek for something behind the
phenomena — they themselves are the theory.

~ *7* ~

Weak minds make the mental error of leaping straight from the particular to the general when, in fact, the general is to be found only within the whole.

~ *8* ~

Nature, however manifold it may appear, is nevertheless always a single entity, a unity; and thus, whenever it manifests itself in part, all the rest must serve as a foundation for the part, and the part must be related to all the rest.

~ *9* ~

What is the universal?
The single case.

What is the particular?
Millions of cases.

~ *10* ~

To grasp that the sky is blue everywhere, one does not have to travel around the world.

~ *11* ~

If you would seek comfort in the whole, you must learn to discover the whole in the smallest part.

~ *12* ~

Nothing is more consonant with Nature than that she puts into operation in the smallest detail that which she intends as a whole.

~ *13* ~

Nothing happens in living Nature that does not bear some relation to the whole ... The question is: how can we find the connection between these phenomena, these events?

~ *14* ~

All effects ... we observe in the world of experience are interrelated in the most constant manner and merge into one another; from first to last they form a series of undulations.

~ *15* ~

The following famous account of Goethe's meeting and recon-ciliation with Schiller in July 1794 (which marked the start of their long and fruitful friendship over the next ten years) was written in 1817. In his account of the meeting, which is here reproduced in full, Goethe describes how he argued for 'another way of con-sidering nature, not piecemeal and isolated, but actively at work, as she proceeds from the whole to the parts.' Intrinsic to this approach is the apprehension *of archetypal forms underlying the multifarious phenomena of nature, which enable the observer to comprehend the latter from the perspective of their deeper unity. Such was the 'archetypal plant' for Goethe: not an abstract idea but a reality experienced on a deeper level. Once grasped, it enables one to perceive the stages of plant growth and metamorphosis as different manifestations of a single protean form.*

PROPITIOUS ENCOUNTER

The happiest moments of my life were experienced during my study of the metamorphosis of plants, as the sequence of their growth gradually became clear to me. This method of regarding the plant world inspired me during my stay at Naples and Sicily; it became more and more precious to me; everywhere I gave myself practice in its application. And these pleasant pursuits were to achieve priceless value by providing the occasion for one of the noblest relationships granted me in my later years. For it was my interest in these phenomena that led to a more intimate association with Schiller, bringing to an end the unfriendly relations that had kept us apart for so long.

On my return from Italy — where I had sought to foster in myself greater precision and purity in all branches of art, unaware of what had meanwhile taken place in Germany — I found certain poetic works, new and old, enjoying great influence and popularity at home, unfortunately of such a sort as to inspire in me the utmost repugnance. I shall mention only Heinse's 'Ardinghello' and Schiller's 'The Robbers.' The first-named author was abhorrent to me because he had used creative art to refine and adorn sensuality and illogical methods of thinking; the second, because a vigorous though immature talent had poured out over the fatherland, in ecstatic torrents, precisely those ethical and dramatic paradoxes of which I had been endeavouring to purify myself.

I did not blame the two men themselves for what they had undertaken and achieved, for no man can gainsay the promptings of his nature. At first he accepts these dictates unconsciously and naively, then more and more consciously at each stage of his development — this being indeed the explanation why so many excellent and foolish things alike are spread over the world and why confusion develops from confusion.

However, the furore these writers had created in my country, the acclaim they were receiving from undisciplined students and

cultivated court ladies alike, appalled me. It seemed to me that my
striving had all been wasted. The pursuits in which I had educated
myself, the manner and style in which I had done it, seemed
shunted aside and invalidated.

The thing that pained me most was that my friends, Heinrich
Meyer and Moritz, and likewise the artists that were working on
the same principles, seemed to be endangered: I felt much concern.
Had it been possible, I would gladly have given up my study of the
plastic arts and the exercise of my literary talent completely. How
could I hope to be heard in competition with those rhapsodical,
extravagant works? Let the reader picture my position. Long
endeavouring to nourish and support the purest conceptions, I now
found myself wedged in between an Ardinghello and a Franz
Moor.

Moritz had also returned from Rome and was spending some
time with me. We lent each other moral support and passionately
fortified our mutual convictions. Schiller, though he lived near me
during his stay in Weimar, I avoided. The appearance of 'Don
Carlos' was not likely to draw us closer together. I resisted the
attempts of mutual friends to bring us together, and thus we
continued to live side by side, yet strangers.

His essay, 'On Grace and Dignity in Literature,' was as little
calculated to conciliate me, for in it he rapturously embraces the
Kantian philosophy, one which elevates the subjective to great
heights while appearing to circumscribe it. The essay plainly
revealed the extraordinary gift Nature had bestowed upon him. Yet
with immoderate feelings of emancipation and self-determination
he played the ingrate to the Good Mother, though she certainly had
not played the stepmother to him. Instead of independently and
actively observing her manner of creating, as she advances according to law from the lowest to the highest forms, he approached her
from the standpoint of a number of empirical human traits. Certain
harsh passages I could even interpret with reference to myself, but
though they showed my convictions in a false light, I felt it would
be even worse if they had not been said with me in mind, for then
the enormous chasm between our modes of thinking must be even
more unbridgeable.

Reconciliation seemed out of the question, and even the mild persuasion of a Dalberg, who had given Schiller the high evaluation he deserved, had no effect. Of course, the reasons I advanced against a reconciliation were hard to refute. No one would deny that in the case of intellectual antipodes more than just the distance of the earth's diameter constitutes the separation, and that, as poles, they cannot actually ever coincide. However, a relation may nevertheless exist between them, as proved by the following incident.

Schiller had moved to Jena, where, as before, I saw nothing of him. It was about this time that Batsch, with unbelievable enterprise, had founded a scientific society, with fine collections and impressive apparatus. I usually attended its periodic meetings and one time found Schiller there. By chance we were left in the hall together, and began a conversation. He appeared to be interested in the lectures, but remarked with great insight, and to my pleasure, that such mangled methods of regarding Nature could only repel a lay person who might otherwise be willing to venture into the subject. I answered that perhaps even to express such a method would be uncongenial and that there might be another way of considering Nature, not piecemeal and isolated but actively at work, as she proceeds from the whole to the parts. Schiller expressed the desire to have the point clarified through discussion, though not concealing his doubts and refusing to grant that my views owed their origins to experience.

We had reached his house; the conversation had lured me in. I gave a spirited explanation of my theory of the metamorphosis of plants with graphic pen sketches of a symbolic plant. He listened and looked with great interest, with unerring comprehension, but when I had ended, he shook his head, saying, 'That is not an empiric experience, it is an idea.' I was taken aback and somewhat irritated, for the disparity in our viewpoints were here sharply delineated. The statement from his essay, 'On Grace and Dignity in Literature,' occurred to me again; the old antipathy was astir. Controlling myself, I replied, 'How splendid that I have ideas without knowing it, and can see them before my very eyes.'

Schiller, who had far greater tact and urbanity than I, who, furthermore, in the hope of procuring my help with his magazine 'The Hours,' was desirous of attracting rather than repelling me, replied in the manner of a trained Kantian. My stubborn realism gave rise to a lively argument, and a great battle ensued. Though later an armistice was called, and neither could consider himself the victor, yet each considered himself invincible. Sentences like the following made me quite unhappy: 'How could any experience ever be gauged by an idea, for the characteristic thing about an idea is that it can never be congruous with an experience.' Yet if he termed an idea what I called an experience, then there must certainly be something negotiable, something in common between us. The first step had after all been taken, and Schiller had great personal magnetism and the power to hold those whom he had attracted. Thus I became interested in his plans and promised to turn over to 'The Hours' some of the manuscripts hidden away in my desk. His wife, whom I had known and loved from her childhood, contributed her share to an enduring friendship, all our mutual friends were happy; and thus through the greatest duel between the objective and the subjective, we sealed a bond which lasted uninterruptedly and accomplished much good for ourselves and others.

After this happy beginning, during ten years of intimate association, such philosophic tendencies as were latent in my nature gradually unfolded. It is my intention to account for this unfolding in so far as it is possible, but the difficulties involved must be strikingly evident to the initiated. Those who command a higher vantage ground from which to survey the easy confidence of the human mind, the understanding innate in the healthy individual with no doubts about facts and their significance nor about his own ability to comprehend, judge, and properly evaluate them — those who command such a vantage point will freely admit that a virtual impossibility is undertaken when one attempts to describe transitions to a purer, freer, and more objective phase — transitions of which there must be thousands upon thousands. We are not speaking here of levels of education, but of those false trails, pitfalls, and circuitous bypaths which are followed by sudden progress and a vigorous upswing to a higher state of culture. What

individual can actually assert that scientifically they always travel
in that highest area of consciousness, where external things can be
observed with the utmost deliberation, with impartiality and vigi-
lance alike, where they work in accord with the law of their inner
nature, with vision and foresight and with the enduring hope of
acquiring a truly pure and harmonious point of view. Does not the
world, do we not ourselves tarnish the lustre of such moments? But
at least we may cherish pious hopes, and a loving attempt to reach
the unattainable is not denied us.

Whatever victories we achieve in the venture, we shall commend
to long-esteemed friends and to the youth of Germany in their
striving toward the good and the right.

May we thus attract and acquire vigorous sympathizers and
future exponents!

~ *16* ~

*Written in 1818, 'Indecision and Surrender' presents the task of
uniting idea and experience as a virtually insurmountable problem
for the scientific researcher. In this respect, the piece is uncha-
racteristic, for in other essays (such as the previous one, and also
the one following this piece), Goethe's viewpoint is quite clear:
namely, that it is possible to gain a 'higher experience' in which
we can have an intuitive perception of the archetypal ideas at work
in phenomena. In such an experience, the division between idea
and experience is annulled. However, we have a more characteristic Goethean theme in the initial statement that 'God is operative
in Nature and Nature in God,' and in the exhortation contained in
the opening lines of the concluding poem to simply 'regard with
silent wonder the Eternal Weaver's masterpiece.'*

INDECISION AND SURRENDER

In observing the cosmic structure from its broadest expanse down
to its minutest parts, we cannot escape the impression that under-

lying the whole is the idea that God is operative in Nature and Nature in God, from eternity to eternity. Intuition, observation, and contemplation lead us closer to these mysteries. We are presumptuous and venture ideas of our own; turning more modest, we merely form concepts that might be analogous to those primordial beginnings.

At this point we encounter a characteristic difficulty — one of which we are not always conscious — namely, that a definite chasm appears to be fixed between idea and experience. Our efforts to overbridge the chasm are forever in vain, but nevertheless we strive eternally to overcome this hiatus with reason, intellect, imagination, faith, emotion, illusion, or — if we are capable of nothing better — with folly.

By honest persistent effort we finally discover that the philosopher might probably be right to assert that no idea can completely coincide with experience, nevertheless admitting that the idea and the experience are analogous, indeed must be so.

In all scientific research the difficulty of uniting idea and experience appears to be a great obstacle, for an idea is independent of time and place but research must be restricted within them. Therefore, in an idea, the simultaneous and successive are intimately bound up together, whereas in an experience they are always separated. Our attempt to imagine an operation of Nature as both simultaneous and successive, as we must in an idea, seems to drive us to the verge of insanity. The intellect cannot picture united what the senses present to it separately, and thus the duel between the perceived and the ideated remains forever unsolved.

For this reason we justifiably take flight into poetry, giving by way of change a new form to an old song:

> Regard with silent wonder
> The Eternal Weaver's masterpiece.
> A single movement sends the shuttle
> Over, under, till the myriad threads
> Meet and interlace, creating
> Countless unions at one stroke!
> The warp, not mounted thread by thread,

But laid down in the timeless past
Awaits the casting of the weft,
Forever waits the Master's will.

~ *17* ~

In this commentary on a passage from Kant's Critique of Judgment, *Goethe seizes on the idea of an intuitive intellectual faculty
that can give access to the universal creative forces within nature.*

In seeking to penetrate Kant's philosophy, or at least apply it as
well as I could, I often got the impression that this good man had
a roguishly ironic way of working: at times he seemed determined
to put the narrowest limits on our ability to know things, and at
times, with a casual gesture, he pointed beyond the limits he
himself had set. He had no doubt observed humanity's precocious
and cocky way of making smug, hurried, thoughtless pronounce-
ments based on one or two facts, of rushing to hasty conclusions
by trying to impose on the objective world some notion that passes
through one's head. Thus our master limits his thinking person to
a reflective, discursive faculty of judgment and absolutely forbids
us one which is determinative. But then, after he has succeeded in
driving us to the wall, to the verge of despair in fact, he makes the
most liberal statements and leaves it to us to decide how to enjoy
the freedom he allows us. In this sense I view the following
passage as particularly significant:

> We can, however, think an understanding which being,
> not like ours, discursive, but intuitive, proceeds from the
> *synthetical-universal* (the intuition of the whole as such)
> to the particular, that is, from the whole to the parts ... It
> is here not at all requisite to prove that such an *intel-
> lectus archetypus* is possible, but only that we are led to
> the idea of it — which too contains no contradiction —
> in contrast to our discursive understanding, which has

need of images *(intellectus ectypus)* and to the contin-
gency of its constitution. *(Critique of Judgment, § 77)*

Here, to be sure, the author seems to point to divine reason. In the
moral area, however, we are expected to ascend to a higher realm
and approach the primal being through faith in God, virtue, and
immortality. Why should it not also hold true in the intellectual
area that through an intuitive perception of eternally creative nature
we may become worthy of participating spiritually in its creative
processes? Impelled from the start by an inner need, I had striven
unconsciously and incessantly towards primal image and prototype,
and had even succeeded in building up a method of representing it
which conformed to nature. Thus there was nothing further to
prevent me from boldly embarking on this 'adventure of reason'
(as the Sage of Königsberg* himself called it).

~ *18* ~

In his *Anthropology* (a book to which we will again refer), Dr
Heinroth† speaks favourably of my work; in fact, he calls my approach unique, for he says that my thinking works objectively.
Here he means that my thinking is not separate from objects; that
the elements of the object, the perceptions of the object, flow into
my thinking and are fully permeated by it; that my perception itself
is a thinking, and my thinking a perception. He does not withhold
his applause for this approach.

In the next few pages I have tried to describe the ideas which
this turn of phrase and the attendant approval aroused in me. I
recommend them to the reader who has looked at the passage in
question *(Anthropology* p.387).

Here, as in earlier volumes, I have followed my intention of

* That is, Kant.

† J.C. Heinroth's *Textbook of Anthropology* was published in 1822, the year
previous to these comments.

describing how I perceive Nature, and also showing something of myself, my inner being, my approach to life. In this regard an earlier essay, 'The Experiment As Mediator between Object and Subject,' will be found especially helpful (see here p.75ff).

I must admit that I have long been suspicious of the great and important-sounding task: 'know thyself.' This has always seemed to me a deception practised by a secret order of priests who wished to confuse humanity with impossible demands, to divert attention from activity in the outer world to some false, inner speculation. Human beings know themselves only in so far as they know the world; they perceive the world only in themselves, and themselves only in the world. Every new object, clearly seen, opens up a new organ of perception in us.

But the greatest help comes from other people: they have the advantage of being able to compare us with the world from their own standpoint, and thus they know us better than we ourselves can.

Since reaching the age of maturity, I have always paid strict attention to what others might know of me: from them and in them, as in so many mirrors, I can gain a clearer idea of myself and what lies within me.

Here I exclude adversaries, for they find my existence odious, repudiate my goals, and condemn my means of reaching them as a mere waste of time. Thus I pass them by and ignore them, for they offer no help with the growth which is the point of life. But friends may call attention to my limitations or to the infinite in my being — in either case I listen to them and trust that they will truly instruct me.

7. The Archetypal Phenomenon

Of all the concepts employed by Goethe in the pursuit of his scientific researches, the concept of the Archetypal Phenomenon (Urphänomen) is the most central. The Archetypal Phenomenon is to be understood as having a status embracing both idea and sensory experience: it is something grasped in conjunction with sensory experience, yet it lies beyond it as a spiritual or ideal quality informing a phenomenon or set of phenomena. The Archetypal Phenomenon is experienced when a group or sequence of phenomena reveal an underlying meaningfulness and internal coherence which is grasped by the intellect in a moment of intuitive comprehension. It is thus revealed to intuitive perception through phenomena, but cannot be reduced to these phenomena alone.

This is not to say that the Archetypal Phenomenon is an abstract concept arrived at by a process of merely selecting out the common elements of a group of phenomena. Goethe refers to it as a 'manifestation': a manifestation of the spiritual ground of phenomena observed by the senses. The intuitive grasp of the Archetypal Phenomenon is a higher kind of experience than what we normally attain to in our observation of nature. It is, in fact, the highest kind of experience a scientist can aim for and, having attained it, one has reached a limit beyond which one should not seek to go. 'When the Archetypal Phenomenon makes one marvel,' Goethe says, 'let one be content.'

~ *1* ~

Written in 1798, 'Experience and Science' is one of a number of short essays on scientific method produced by Goethe during the early years of his friendship with Schiller.

The method here outlined is strictly phenomenological, the intention being to avoid channelling what is observed into the confines of a favoured theory, but rather to lay bare a 'pure and constant phenomenon' underlying and linking together the mass of data given to experience. Unlike a conceptual explanation or theory, the 'pure phenomenon' is something which can be perceived in and through the mass of empirical data when one has sufficiently steeped oneself in them. The 'pure phenomenon' (das reine Phänomen) is Goethe's earlier term for what he later referred to as the 'Archetypal Phenomenon' (Urphänomen).

EXPERIENCE AND SCIENCE

Phenomena, also called facts in lay language, are certain and definite by nature, but often indefinite and variable as they meet the eye. The scientist attempts to grasp and hold fast what is definite in the phenomena; in individual cases he is concerned not only with their actual but also with their ideal appearance. As I have occasion to notice in my present field of work, empirical breaks must often be disregarded in order to preserve a pure, constant phenomenon. However, as soon as I permit myself to do this, I am establishing a kind of ideal.

Nevertheless, a vast difference exists between disregarding whole sequences in favour of a hypothesis, as theorists often do, and the sacrifice of a single empirical break in the interest of preserving the idea of the pure phenomenon.

Since we, therefore, as observers never see pure phenomena with our eyes, since much depends instead upon our own state of mind, on the state of the organ itself at the moment, on light, air, weather, bodies, treatment, and a thousand other things, it would be like attempting to drink up the ocean if we were to fasten upon each and every phenomenon with the intention of observing, measuring, judging, and describing them individually.

In my observation and contemplation of Nature, especially of late, I have remained as faithful as possible to the following method.

After observing a certain degree of constancy and logical sequence in phenomena, I derive an empirical law and prescribe it for future phenomena. If the law and the phenomena later coincide completely, I am vindicated; if they do not, at least my attention has been drawn to the details of individual cases and to the necessity for more correct methods of organizing the contradictory experiments. However, if under similar circumstances, a case appears that contradicts my law, I know I must pass on and search for a higher viewpoint.

According to my experience, this then would be the point where the human mind could approach closest to the objects in their general aspects, absorb them, and in rational terms amalgamate with them (as we also do at the practical level).

The object of our work would then be to demonstrate: (1) the empirical phenomenon, of which every individual is conscious in Nature and which later is elevated to (2) a scientific phenomenon by experimentation, by representing it under circumstances and conditions differing from those in which we first encountered it, and in a more or less effective sequence; and (3) the pure phenomenon now standing forth as the result of all experiences and experiments. It can never be isolated, appearing as it does in a constant succession of forms. In order to describe it, the human intellect determines the empirically variable, excludes the accidental, separates the impure, unravels the tangled, and even discovers the unknown.

Here we would reach the ultimate goal of our powers, if human beings knew their place. For we are not seeking causes but the circumstances under which the phenomenon occurs. Its logical sequence, its eternal return under a thousand conditions, its uniformity and mutability are considered and accepted; its definiteness is recognized and redefined by the human intellect. And in my opinion such work is certainly not mere speculation, but rather the practical and self-correcting operation of ordinary common sense as it ventures out into a higher sphere.

~ 2 ~

The Archetypal Phenomenon *(Urphänomen)* is not to be regarded as a basic theorem leading to a variety of consequences, but rather as a basic manifestation enveloping the specifications of form for the beholder. Contemplation, knowledge, divination, faith — all these feelers with which human beings reach out into the universe must set to work jointly if we are to fulfil our important but difficult task.

~ 3 ~

In general, events we become aware of through experience are simply those we can categorize empirically after some observation. These empirical categories may be further subsumed under scientific categories leading to even higher levels. In the process we become familiar with certain requisite conditions for what is manifesting itself. From this point everything gradually falls into place under higher principles and laws revealed not to our reason through words and hypotheses, but to our intuitive perception through phenomena. We call these phenomena *archetypal pheno-mena* because nothing higher manifests itself in the world; such phenomena, on the other hand, make it possible for us to descend, just as we ascend, by going step by step from the archetypal phenomena to the most mundane occurrence in our daily experience.

~ *4* ~

But even where we find such an archetypal phenomenon, a further
problem arises when we refuse to recognize it as such, when we
seek something more behind it and above it despite the fact that
this is where we ought to acknowledge the limit of our perception.
It is proper for the natural scientist to leave the archetypal phenomenon undisturbed in its eternal repose and grandeur, and for the
philosopher to accept it into his realm. There he will discover that
a material worthy of further thought and work has been given him,
not in individual cases, general categories, opinions and hypotheses,
but in the basic and archetypal phenomenon.

~ *5* ~

Beauty is an archetypal phenomenon. While it never materializes
as such, it sheds its glow over a thousand different manifestations
of the creative spirit and is as multiform as Nature itself.

~ *6* ~

Archetypal phenomena: ideal, real, symbolic, identical. Empirical
realm: endless proliferation of these, thus hope of succour, despair
of perfection.

> Archetypal phenomenon:
> ideal as the ultimate we can know,
> real as what we know,
> symbolic, because it includes all instances,
> identical with all instances.

~ 7 ~

The direct experience of archetypal phenomena creates a kind of anxiety in us, for we feel inadequate. We enjoy these phenomena only when they are brought to life through their eternal interplay in the empirical.

~ 8 ~

When archetypal phenomena stand unveiled before our senses we become nervous, even anxious. Sense-based people seek salvation in astonishment, but soon that busy matchmaker, Understanding *(Verstand)*, arrives with her efforts to marry the highest to the lowliest.

~ 9 ~

The highest thing a person can attain to is to marvel. When the Archetypal Phenomenon makes one marvel, let one be content. It cannot afford one an experience beyond this, and to seek something else behind it is futile. Here is the limit. But as a rule people are not satisfied to behold an Archetypal Phenomenon. They think there must be something beyond. They are like children who, having looked into a mirror, turn it around to see what is on the other side.

~ 10 ~

If ultimately I rest content with the Archetypal Phenomenon, it is, after all, but a kind of resignation; yet it makes a great difference whether I resign myself at the boundaries of humanity, or within a hypothetical narrowness of my small-minded individuality.

~ *11* ~

The simple archetype I hit on long ago. No organic being wholly corresponds to the underlying idea. The higher idea lurks behind each. That is my God; that is the God we all seek and hope to set our countenance upon; but we can only divine him, not see him.

~ *12* ~

... Nature understands no jesting; she is always true, always serious, always severe; she is always right, and the errors and faults are always those of the human being. The person incapable of appreciating her she despises; and only to the apt, the pure, and the true, does she resign herself, and reveal her secrets.

The Understanding *(Verstand)* will not reach her; one must be capable of elevating oneself to the highest Reason *(Vernunft)* to come into contact with the Divine, which manifests itself in the archetypal phenomena which dwell behind them and from which they proceed.

8. God in Nature, Nature in God

The experience of the Archetypal Phenomenon (Urphänomen) *was, for Goethe, a spiritual experience. Not a spiritual experience as distinct from a sensory experience, but rather* in conjunction with *a sensory experience. As well as such special moments in which the Archetypal Phenomenon is revealed, nature constantly offers, to one who is receptive to it, intimations of the divine. For Goethe, the divine is present everywhere in nature. This was not simply a belief, nor an article of faith; it was an intrinsic part of his experience of nature, as the quotations in this chapter show.*

~ *1* ~

Jacobi's book *On Divine Matters* made me feel ill at ease. Dearly beloved friend that he was, how could I welcome the development of the thesis that Nature conceals God? My own pure, deep, innate and schooled view of things had taught me without fail to see God in Nature, Nature in God, and this view was the foundation of my very existence.

~ *2* ~

Forgive me for preferring to keep silent when you talk of a divine being. I discern such a one only in and by means of the *res singulares*. To the closer and deeper study of these no one has a happier faculty of leading than Spinoza himself, despite the fact that under his gaze all individual things seem to disappear. Here I am on and among mountains, seeking the divine in *herbis et lapidibus*.

~ *3* ~

'Nature conceals God!' But not from everyone.

~ *4* ~

As if the external world did not everywhere reveal to those who have eyes the most mysterious laws by day and night. In this persistence of the infinitely manifold I *see* most clearly the hand-writing of God.

~ *5* ~

When you say that one can only *believe* in God ... I reply that I rely more on *seeing*. And when Spinoza speaks of 'intuitive knowledge,' and says: 'This species of knowing proceeds from an adequate idea of the formed essence of certain attributes of God to the adequate knowledge of the essence of things': these few words give me the courage to devote my whole life to the contemplation of things which are within my reach, and of whose formal essence I can hope to construct an adequate idea.

~ 6 ~

Assuredly, there is no more lovely worship of God than that for which no image is required, but which springs up in our breast spontaneously, when Nature speaks to the soul, and the soul speaks to Nature face to face.

~ 7 ~

'Nature does nothing in vain,' is an old Philistine maxim. She works ever vitally, superfluously, and lavishly, so that the Infinite may be ever present, because nothing can last.

~ 8 ~

Though You hide yourself in a thousand forms,
Yet, most beloved, at once I recognize You;
Though You cover yourself in a thousand magic veils,
Yet, ever present, at once I recognize You.

In tausend Formen magst du dich verstecken,
Doch, Allerliebste, gleich erkenn ich dich;
Du magst mit Zauberschleiern dich bedecken,
Allgegenwärtge, gleich erkenn ich dich.

~ 9 ~

In observing the cosmic structure from its broadest expanse down to its minutest parts, we cannot escape the impression that underlying the whole is the idea that God is operative in Nature and Nature in God, from eternity to eternity.

~ *10* ~

And an eternal, living Activity
works to create anew what has been created
lest it entrench itself in rigidity.
And that which has not yet been
seeks now to come into being
as pure suns and many-coloured earths:
none of it may remain at rest.

It is intended to move, to act and create —
first to form and then to transform itself;
its moments of immobility are only apparent.
In all that lives the Eternal Force works on;
for everything must dissolve into nothingness,
if it is to remain in Being.

Und umzuschaffen das Geschaffne
Damit sichs nicht zum Starren waffne,
Wirkt ewiges, lebendiges Tun.
Und was nicht war, nun will es werden
Zu reinen Sonnen, farbigen Erden;
In Keinem Falle darf es ruhn.

Es soll sich regen, schaffend handeln,
Erst sich gestalten, dann verwandeln;
Nur scheinbar stehts Momente still.
Das Ewige regt sich fort in allen;
Denn alles muss in Nichts zerfallen,
Wenn es im Sein beharren will.

~ *11* ~

Nothing that is can dissolve into nothingness!
In all that lives the Eternal Force works on:
remain, rejoicing, in Being!
Being is eternal;
for laws preserve the living treasures
with which the universe has adorned herself.

Kein Wesen kann zu Nichts zerfallen!
Das Ewige regt sich fort in allen,
Am Sein erhalte dich beglückt!
Das Sein ist ewig: denn Gesetze
Bewahren die lebend'gen Schätze,
Aus welchen sich das All geschmückt.

~ *12* ~

The works of Nature are like a freshly spoken word of God.

9. Seeing and Seeing

One of the most important and distinctive features of Goethe's scientific method is his insistence that, in the quest for scientific knowledge, none of the human faculties should be excluded. Essays such as 'The Experiment as Mediator Between Subject and Object' show how deeply Goethe was aware of the importance of guarding against fantasy, wish-fulfilment and other subjective projections in scientific work. But this did not lead him to advocate a scientific method based purely on empirical analysis, as the only guarantor of objectivity. On the contrary, just as observation and thinking must be disciplined so that they accurately reflect what exists in the world, so also must our other faculties. An object investigated by empirical analysis alone will reveal only as much of itself as the limitations of this method will permit. If we allow intuitive thinking, feeling and imagination a place in our scientific method, then — providing these are deployed in conjunction with exact observation and clear thought, and providing they are trained as thoroughly as our powers of observation and thinking — then a much fuller and more complete experience of nature will become possible.

~ *1* ~

... there is a difference between seeing and seeing; ... the eyes of the spirit have to work in perpetual living connection with those of the body, for one otherwise risks seeing and yet seeing past a thing.

~ *2* ~

Every new object, clearly seen, opens up a new organ of perception
in us.

~ *3* ~

None of the human faculties should be excluded from scientific
activity. The depths of intuition *(Ahnung)*, a sure awareness of the
present, mathematical profundity, physical exactitude, the heights
of reason *(Vernunft)* and sharpness of intellect *(Verstand)* together
with a versatile and ardent imagination, and a loving delight in the
world of the senses — they are all essential for a lively and
productive apprehension of the moment.

Even if these required elements may seem to be, where not
contradictory, yet nevertheless opposed to one another in such a
way that even the most excellent spirits could not hope to unite
them: they nevertheless still reside manifestly in the whole of
humanity and can emerge at any moment, as long as they are not
(through prejudices, through the wilfulness of the individuals who
possess them, or through any of the other mistaking, terrifying, and
exterminating negatives there may be) suppressed in the very
moment when alone they can be effective, and their appearance
annihilated in genesis.

~ *4* ~

Although it is true that colours and light are intimately related to
one another, we must consider both as belonging to all Nature.
Through them Nature in her entirety seeks to manifest herself, in
this case to the sense of sight, to the eye.

Similarly, the whole of Nature reveals herself to yet another
sense. Let us shut our eyes, let us open our ears and sharpen our
sense of hearing. From the softest breath to the most savage noise,

from the simplest tone to the most sublime harmony, from the fiercest cry of passion to the gentlest word of reason, it is Nature alone that speaks, revealing her existence, energy, life and interrelations, so that a blind man to whom the vast world of the visible is denied may seize hold of an infinite living realm through what he can hear.

Thus Nature also speaks to other senses which lie even deeper, to known, unrecognized and unknown senses. Thus she converses with herself and with us through a thousand phenomena. No one who is attentive will ever find Nature dead or silent.

~ 5 ~

In the human spirit, as in the universe, nothing is higher or lower; everything has equal rights to a common centre which manifests its hidden existence precisely through this harmonic relationship between every part and itself. The quarrels in antiquity, as well as in modern times, all spring from a division of what God created as one in His realm of Nature. We are well enough aware that some skill, some ability, usually predominates in the character of each human being. This leads necessarily to one-sided thinking since we know the world only through ourselves, and thus have the naïve arrogance to believe that the world is constructed by us and for our sake. It follows that we put our special skills in the foreground, while seeking to reject those we lack, to banish them from our own totality. As a correction, we need to develop all the manifestations of human character — sensuality and reason, imagination and common sense — into a coherent whole, no matter which quality predominates in us. If we fail to do so, we will labour on under our painful limitations without ever understanding why we have so many stubborn enemies, why we sometimes meet even ourselves as an enemy.

Thus someone born and bred to the so-called exact sciences, and at the height of their ability to reason empirically, finds it hard to accept that an exact sensory imagination might also exist, although art is unthinkable without it. This is also a point of contention between followers of emotional religion and those of rational religion: while the latter refuse to acknowledge that religion begins with feeling, the former will not admit the necessity for religion to develop rationally.

~ 6 ~

A great scientist without this high gift [of imagination] is impossible. I do not mean an imagination that goes into the vague and imagines things that do not exist; I mean one that does not abandon the actual soil of the earth, and steps to supposed and conjectured things by the standard of the real and the known. Then it may prove whether this or that supposition be possible, and whether it is not in contradiction with known laws. Such an imagination presupposes an enlarged tranquil mind, which has at its command a wide survey of the living world and its laws.

~ 7 ~

Imagination is first re-creative, repeating only the objects. Furthermore, it is productive by animating, developing, extending, transforming the objects. In addition, we can postulate a perceptive imagination which apprehends identities and similarities ... Here it becomes evident how desirable analogy is which carries the mind to many related points, so that its activity can unite again the homogeneous and the homologous.

~ *8* ~

There is a great difference between dimly apprehending with dull senses a vague whole, and seeing and grasping directly a complete whole.

~ *9* ~

When, having something before me that has grown, I enquire after its genesis and measure the process as far back as I can, I become aware of a series of stages that, although I cannot actually see them in succession, I can present them to myself in memory as a kind of ideal whole.

~ *10* ~

(a) I had the ability, with my eyes closed and my head lowered, to evoke the image of a flower in the centre of my organ of visualization; and to perceive the flower in such a way that it did not remain in its original form for a single moment, but spread out, and from within there unfolded again new flowers with coloured as well as green leaves. They were not natural flowers by any means, but products of the imagination, albeit as regularly shaped as stonemason's rosettes ... It did not occur to me to experiment like this with other objects. Perhaps these offered themselves so readily because they had their roots in many years of contemplation of the metamorphosis of plants.

(b) Here the phenomenon of the after-image *(Nachbild)*, memory, productive imagination, concept and idea are all in play at once, and manifest themselves within the living dynamic of the organ with complete freedom, and without purpose or guidance.

~ *11* ~

Intuitive perception *(Anschauung)* gives us at once the complete concept of an achieved form; the faculty of thought *(Denk-Kraft)*, not wishing to lag behind, shows and articulates in its own way how such a form could and must be achieved ...

Since [the faculty of thought] does not feel entirely adequate to the task, it calls on the imagination for help; and in this way conceptual entities *(entia rationis)* gradually arise whose great advantage it is to lead us back to perception and to pressure us into greater attention and complete insight.

~ *12* ~

To grasp the phenomena, to fix them to experiments, to arrange the experiences and know the possible modes of representation of them — the first as attentively, the second as accurately, the third as exhaustively as possible and the last with sufficient many-sidedness — demands a moulding of man's poor ego, a transformation so great that I never should have believed it possible.

~ *13* ~

Everything we call invention and discovery in the higher sense is ... the activation of an internal feeling for truth, which has long unobtrusively been developing and then suddenly and unexpectedly leads with lightning speed to fruitful knowledge. It is a revelation which proceeds from within the human being, to act on the external world, and it gives us an intimation of our kinship with God. It is a synthesis of mind with the external world, and it gives the most comforting assurance of the eternal harmony of all that exists.

10. The Horizon Boundless

In the last chapter, we have seen Goethe striving to push back the limits to human knowledge. If the infinite and eternal divine life manifests in the finite and temporal world, and can be perceived there by human beings whose lives are bounded by space and time, then the question arises as to how far we must accept the apparent limitations of scientific knowledge of the natural world, and how far this may be extended into the spiritual.

The quotations gathered together in this chapter are metaphysical reflections concerning the relationship of subject to object, of humanity to God, of finite to infinite, and of the knowable to the unknowable. In general, Goethe's attitude is that, while accepting that in practice there are boundaries to human knowledge, and that nature herself has an unfathomable dimension, the onus is on us to press these boundaries back as far as possible.

~ *1* ~

Written in 1820, 'A Friendly Greeting' states the important principle that the depth and extent of human knowledge has no bounds. This is in contrast to the epistemology of Kant's Critique of Pure Reason *in which the restrictions to human knowledge are carefully defined, and preclude us from ever having the possibility of knowing 'things in themselves.' The Kantian maxim quoted in the poem 'Into the core of Nature/No Earthly mind can enter' is from Albrecht von Haller's 'The Falsehood of Human Virtues' (1730) — a popular didactic poem of the eighteenth century.*

A FRIENDLY GREETING

I can no longer conceal a pleasure which has come upon me more than once in recent days. I have a wonderful feeling of being in harmony with serious, productive researchers here and elsewhere. Although they admit the need to postulate and acknowledge something beyond knowing, they do not draw a line the researcher himself is forbidden to cross.

Must I not acknowledge and postulate myself without ever knowing my own nature? Do I not endlessly study myself without ever achieving a grasp of myself, or myself and others? Yet we cheerfully continue to press forward.

The same is true of the world! It may lie before us without beginning or end, the horizon boundless, our surroundings impenetrable — so be it. No limit, no definition, may restrict the range of depth of the human spirit's passage into its own secrets or the world's. The following bit of light verse should be read and understood in this spirit:

Spontaneous Outburst

'Into the core of Nature'
— O Philistine —
'No earthly mind can enter.'

The maxim is fine;
But have the grace
To spare the dissenter,
Me and my kind.
We think: in every place
We're at the centre.

'Happy the mortal creature
To whom she shows no more
Than the outer rind,'
For sixty years
I've heard your sort announce.

It makes me swear, though quietly;
To myself a thousand times I say:
All things she grants, gladly and lavishly;
Nature has neither core
Nor outer rind,
Being all things at once.
It's yourself you should scrutinize to see
Whether you're centre or periphery.

~ 2 ~

Search within yourselves and you will find everything; and be glad
if out there, however you choose to name it, a nature lies which
says 'Yes and Amen' to everything that you have found within
yourselves.

~ 3 ~

There is a secret element of regularity in the object which corresponds to a secret element of regularity in the subject.

~ 4 ~

Had I not harboured the world within me by anticipation, I would
have remained blind with seeing eyes, and all research and experience would have been a lifeless and futile effort.

~ 5 ~

Where object and subject touch, there is life.

~ 6 ~

Science of nature has one goal:
To find both manyness and whole.
Nothing 'inside' or 'Out There,'
The 'outer' world is all 'In Here.'
This mystery grasp without delay,
This secret always on display.
The true illusion celebrate,
Be joyful in the serious game!
No living thing lives separate:
One and Many are the same.

Müsset im Naturbetrachten
Immer eins wie alles achten:
Nichts ist drinnen, nichts ist draussen;
Denn was innen, das ist aussen.
So ergreifet, ohne Säumnis,
Heilig öffentlich Geheimnis.

Freuet euch des wahren Scheins,
Euch des ernsten Spieles:
Kein Lebendiges ist ein Eins,
Immer ists ein Vieles.

~ 7 ~

To know Nature, one ought to be nature itself. What one is able to express of nature is always something specific, that is, it is something real, something actual, namely something in relation to oneself. But what we express is not all that is; it is not its whole nature. This may serve as an explanation, and concession to those who still speak of things-in-themselves. Although they can say nothing of things-in-themselves just because they *are* things-in-themselves, that is, are out of relation to us and we to them, and because we recognize everything that we say to be our own mode of representation ... it is evident that they at least agree with us that what human beings can predicate of things does not exhaust their nature, that they are not simply and solely what they are thus said to be, but much more, and much else ... In other words, things are infinite in their natures. The human being, in expressing the object, is below and above it, humanity and God reconciled in one nature. We should not speak of things-in-themselves, but rather of the One-in-itself. For 'things' exist only from the human point of view, which posits a diversity and a multiplicity. All is actually only one, but who is able to speak of this One as it is in itself?

~ *8* ~

One cannot properly speak of many problems in the natural sciences if one does not draw on metaphysics for help; but not that school-and-word wisdom; rather that which was, is and shall be before, with and after physics.

~ *9* ~

The concepts of being and totality are one and the same; when pursuing the concept as far as possible, we say that we are conceiving of the infinite. But we cannot think of the infinite, or of total existence. We can conceive only of things which are finite or made finite by our mind, that is, the infinite is conceivable only in so far as we can imagine total existence — but this task lies beyond the power of the finite mind.

~ 10 ~

Although all finite beings exist within the infinite, they are not
parts of the infinite; instead, they partake of the infinite.

~ 11 ~

The things we call the parts in every living being are so
inseparable from the whole that they may be understood only in
and with the whole.

... A finite living being partakes of infinity, or rather, it has
something infinite within itself. We might better say: in a finite
living being the concepts of existence and totality elude our
understanding: therefore we must say that it is infinite, just as we
say that the vast whole containing all beings is infinite.

~ 12 ~

My desire to see the formless formed, the infinite arrayed in
regular sequence of form, follows from all my work in science and
art.

~ 13 ~

All of existence and totality must be made finite in our minds so
that it conforms to our nature and our way of thinking and feeling.
Only then will we say that we understand something, or enjoy it.

~ *14* ~

We can never directly see what is true, that is, identical with what is divine: we look at it only in reflection, in example, in the symbol, in individual and related phenomena. We perceive it as a life beyond our grasp, yet we cannot deny our need to grasp it.

~ *15* ~

The mind may perceive the seed, so to speak, of a relation which would have a harmony beyond the mind's power to comprehend or experience once the relation is fully developed. When this happens, we call the impression sublime; it is the most wonderful bestowed on the mind of man.

~ *16* ~

I shall tell you something that may serve as a guide in life: There is in Nature what is within reach and what is beyond reach. Ponder this well and with respect. A great deal is already gained if we impress this general fact upon our mind, even though it always remains difficult to see where the one ends and the other begins. Those who are unaware of the distinction may waste themselves in lifelong toil trying to get at the inaccessible without ever getting close to truth. But those who know it and are wise will stick to what is accessible; and in exploring this region in all directions and confirming their gains they will even push back the confines of the inaccessible. Even so they will have to admit in the end that some things can be mastered only to a certain degree and that Nature always retains a problematic aspect too deep for human faculties to fathom.

~ *17* ~

It is my opinion that humanity must indeed assume an unknowable element, but that we should set no limits to our quest. For even though Nature has the better of us, seeming to keep many of her secrets from us, we have an advantage of our own in that our thoughts may soar beyond Nature while not yet fully comprehending her. We go far enough when we come to the archetypal phenomena, seeing them face to face in their unknowable glory and then turning back to the world of other phenomena. The incomprehensible, in its simplicity, manifests itself in thousands upon thousands of variations, unchanged despite its inconstancy.

~ *18* ~

The highest achievement of the human being as a thinking being is to have probed what is knowable and quietly to revere what is unknowable.

Sources and bibliography

German sources

There is a large array of editions of Goethe's works. Quotations have been taken from different editions, which are indicated by the following abbreviations:

LA   *Goethe: Die Schriften zur Naturwissenschaft* edited by G. Schmidt *et al.,* in *Aufrage der Deutschen Akadamie der Naturforscher* (Leopoldina) Weimar, 1947 ff.

WA   *Goethes Werke,* Weimarer Ausgabe, Weimar, 1887–1919. Four divisions, 143 volumes.

JA   *Goethes Sämtliche Werke,* Jubilaums-Ausgabe; Stuttgart and Berlin: J.G. Cotta'sche Buchhandlung Nachfolger, 1902–7. Forty volumes.

HA   *Goethes Werke,* Hamburger Ausgabe; edited by Erich Trunz *et al.,* Hamburg, Wegner 1948–60. Fourteen volumes.

GnS   *Goethes naturwissenschaftliche Schriften,* edited by Rudolf Steiner, Union Deutsche Verlagsgesellschaft, Stuttgart-Berlin-Leipzig, 1884–97 (1921). Four volumes.

MR   *Maxims and Reflections* in Volume 12 of H.A.

MuR   *Goethe: Maximen und Reflexionem,* edited by Max Hecker, Weimar, 1907 (Schriften der Goethe-Gesellschaft, 21).

G.G.   *Goethes Gespräche,* edited by Flodoard Freiherr von Biedermann, Leipzig 1909–11. Five volumes.

GWBG   *Goethe: Gedenkausgabe der Werke, Briefe und Gespräche,* edited by E.Beutler, Zurich 1948–60.

D.L.   *Goethe: Selected Verse,* German text edited and translated by David Luke, Penguin 1964.

In each case, the reference is given by volume number (where there is more than one volume) followed by page number. Where there are three numbers, the first refers to the part. Hence, for example, both LA and WA references are by part, volume and page.

There are, however, some exceptions to this rule. References to Goethe's

Maxims and Reflections are given as MR or MuR followed by the number of the maxim or reflection, not the page number.

Letters are indicated by their date and the person to whom they were sent. The German may be referred to in *Goethes Briefe Hamburger Ausgabe* (Christian Wegner Verlag, Hamburg 1962. Four volumes) or in WA part IV.

Conversations with Eckermann are also given simply with a reference to date. There are so many German editions of this popular work that it seems unnecessary to pick out any single one.

Similarly, references to Goethe's diary are given only with the date. The German may be referred to in WA I.35 (1749 –1806) and WA I.36 (1807–22).

Sources of translation

I have used the following sources, here listed alphabetically, for translations of Goethe's writings. Some translations have been slightly adapted for stylistic reasons. Where no source of translation is given, the translation is my own.

Amrine, F. (ed.), *Goethe and the Sciences: A Reappraisal* (Dordrecht, D. Reidel Publishing Co., 1987).

Blackie, J.S. (ed.), *The Wisdom of Goethe* (London, 1883).

Coleridge, A.D. (ed.), *Goethe's Letters to Zelter* (London, 1887).

Eastlake, C.L., *Goethe's Theory of Colours* (London, 1840; Cambridge, Mass., MIT Press, 1970).

Gray, R., *Goethe: A Critical Introduction* (Cambridge University Press, 1967).

Heinemann, F., 'Goethe's Phenomenological Method' in *Philosophy* IX.33, Jan., 1934.

Lehrs, E., *Man or Matter* (London, Faber and Faber, 1951).

Ludwig., E., *The Practical Wisdom of Goethe* (London, Allen and Unwin, 1935).

Luke, D. and Pick, R., *Goethe: Conversations and Encounters* (London, Oswald Wolff, 1966).

Luke, D., *Goethe: Selected Verses* (Penguin, 1964).

Magnus, R., *Goethe as a Scientist,* trans. H. Norden (New York, Henry Schuman, 1949).

Miller, D., *Goethe: Scientific Studies* (New York, Suhrkamp, 1988. New edition, Princeton University Press, 1995).

Mueller, B., *Goethe's Botanical Writings* (Honolulu, University of Hawaii Press, 1952).

Nisbet, H.B., *Goethe and the Scientific Tradition* (London, Institute of Germanic Studies, 1972).

Oxenford, J., *Conversations of Goethe with Eckermann* (London, Everyman, 1930).

Roszak, T., *Where the Wasteland Ends* (London, Faber, 1972).

Sepper, D., *Goethe contra Newton* (Cambridge University Press,1988).

Steiner, R., *Goethe the Scientist,* trans. O.D. Wannamaker (New York, Anthroposophic Press, 1950).

Stephenson, R.H., *Goethe's Wisdom Literature* (New York, Peter Lang, 1983).

—, 'Last Universal Man — or Wilful Amateur?' in Wilkinson, E.M., (ed.) *Goethe Revisited* (London, John Calder, 1984).

Vasco, G.M., *Diderot and Goethe* (Geneva, Librairie Slatkine, 1978).

Vietor, K., *Goethe the Thinker* (Mass., Harvard University Press, 1950).

Weigand, H.J., (ed.) *Goethe: Wisdom and Experience* (London, Routledge and Kegan Paul, 1949).

Wells, G.A., *Goethe and the Development of Science, 1750–1900* (Sijhoff and Noordhoff, 1978).

Key to sources

Title page quote from Goethe's review of *Notes for a Physiognomy of Plants* by Alexandre von Humboldt, 1806. Trans. Mueller p.122.

1. THE HUMAN BEING IS THE MOST EXACT INSTRUMENT

1a.   MR 664 trans. Miller p.311 (adapted).

1b.   MR 665 trans. Collison p.55.

1c.   MR 666 trans. Collison p.55 (adapted).

2.   from *'Vermächtnis'* D.L. 277 trans. Luke.

3.   JA 4: 242 trans. Heinemann p.76.

4.   MuR 1193 trans. Stephenson 1983 p.92.

5.   MuR 502 trans. Stephenson 1983 p.118.

6.   *Wilhelm Meisters Wanderjahre,* JA 19.139 trans. Vietor p.42 (adapted).

7.   MR 617 trans. Miller p.309.

8.   MR 620 trans. Miller p.309.

9.   MR 569 trans. Miller p,308.

10.   MR 573 trans. Miller p.308.

11.   MuR 1072 trans. Weigand p.125.

12.   *Theory of Colour (Zur Farbenlehre* 1810) Historical Part; Div.5. Galileo Galilei. WA II.3.247; trans. Weigand p.118.

13.   *Theory of Colour* Introduction. HA 13.322–23; trans. Miller (adapted) p.163.

14.   *Theory of Colour* WA II.1; trans. Eastlake p.283.

15.   *Theory of Colour* Part 5 section 751. HA 13.491–92; trans. Miller p.277.

16.   *Theory of Colour* Preface. HA 13.315; trans. Miller p.158 (adapted).

17.   from 'A Good Proposal' *(Verschlag zur Güte)* WA II.11.65; trans. Stephenson 1984 p.70.

18.   Conversations with Eckermann 18/5/1824 trans. Oxenford pp.65–66

(adapted).

2. OBSERVATION OF NATURE IS LIMITLESS

1. from 'Problems' *(Probleme,* 1823) HA 13.35; trans. Miller p.43.
2. from 'Gott, Gemüt und Welt' D.L. 280.
3. from 'Preliminary Notes for a Physiology of Plants' *(Vorarbeiten zu einer Physiologie der Pflanzen)* WA II. 6.301–3; trans. Mueller p.92 (adapted).
4. *Theory of Colour (Zur Farbenlehre)* V sections 751–54. HA 13.491–92; trans. Miller p.277 (adapted).
5. MR 709 trans. Miller p.312.
6. from 'My Relationship to Science, and to Geology in Particular' *(Verhältnis zur Wissenschaft, Besonders zur Geologie, 1820)* HA 13.273; trans. Miller p.138–39.
7. from 'Preliminary Notes for a Physiology of Plants' *(Vorarbeiten zu einer Physiologie der Pflanzen)* WA II.6.289–92; trans. Mueller pp.86–88.
8. from 'Outline for a General Introduction to Comparative Anatomy' *(Erster Entwurf einer allgemeinen in die vergleichende Anatomie, ausgehend von der Osteologie, 1820)* HA 13.170; trans. Miller p.117.
9. MR 662 trans. Miller p.311.
10. MR 399 trans. Miller pp.304–5.
11a. MR 407 trans. Miller p.305.
11b. MR 408 trans. Miller p.305.
12. Letter to Jacobi 6/1/1813 trans. Weigand pp.45–46.
13. MR 444 trans. Miller p.306 (adapted).

3. SETTING FORTH A MORPHOLOGY

1. From 'The Purpose is Set Forth' *(Die Absicht eingeleitet,* 1817) HA 13.54–56; trans. Miller pp.63–64.
2. Diary 22/4/1812.
3. Conversations with Eckermann 13/2/1829 trans. Weigand p.75–76.
4a. MR 595 trans. Miller p.309.
4b. MR 596 trans. Miller p.309.
5. from 'The Formative Impulse' *(Bildungstrieb,* 1820) HA 13.33–34; trans. Miller p.36.
6. MR 21 trans. Miller pp.303–4.
7. from 'Polarity' *(Polaritat,* 1799) WA II.11.165; trans. Miller p.155–56.
8. *Theory of Colour (Zur Farbenlehre)* Didactic Part, JA 40. 83; trans. Weigand p.128.
9. 'Analysis and Synthesis' *(Analyse und Synthese)* HA 13.49–52; Trans. Miller pp.48–50.
10. From second review of E. Geoffroy St. Hilaire's *Principes de Philosophie Zoologique* (1832) JA 39.233; trans. Weigand p.131.
11. MuR 561 trans. Stephenson 1983 p.112 (adapted).

12. from Campaign in France, Pempelfort, Nov. 1792, JA 28.155 trans. Weigand p.129.
13. Letter to von Muller 24/5/1828 HA 13.48 trans. Miller p.6.
14. from 'Our Objective is Stated' *(Die Absicht eingeleitet, 1817)* HA 13.56; trans. Mueller p.24.
15. from *Metamorphosis of Plants,* 2nd Essay. WA II.6.282; trans. Mueller p.80.
16. from 'On the Spiral Tendency in Plants' *(Spiraltendenz der Vegetation)* HA 13.134; trans. Vietor p.56.
17. from 'God, Heart and World' *(Gott, Gemüt und Welt)* WA I.2.216; trans. Nisbet p.20.
18. from 'The Experiment as Mediator' *(Der Versuch als Vermittler, 1823)* para 24. HA 13.17; trans. Miller p.15.
19. from 'Toward a Theory of Weather' *(Versuch einer Witterungslehre, 1825)* HA 13.306–7; trans. Miller p.145–46.
20. Conversations with Riemar 19/3/1807 GWBG 22.440; trans. Luke and Pick p.67.
21. from 'Toward a General Comparative Theory' *(Versuch einer allgemeinen Vergleichungslehre)* LA I.10.118–22; trans. Miller pp.54–56.

4. QUANTITY AND QUALITY: TWO POLES OF MATERIAL EXISTENCE

1. MR 641 trans. Miller p.309–10.
2. MR 644 trans. Miller p.310 (adapted).
3. MR 643 trans. Miller p.310 (adapted).
4. MuR 1280 trans. Stephenson 1983 p.104 (adapted).
5. from 'Tibia und Fibula' (1824) WA II.8.219; trans. Nisbet p.49.
6. Conversations with Eckermann 20/12/1826 trans. Oxenford p.139.
7. from 'Ferneres Über Mathematik und Mathematiker' WA II. 11.98; trans. Heinemann p.69.
8. from 'A Study on Spinoza' *(Studie nach Spinoza,* c.1793) HA 13.7; trans. Miller p.8.
9. *Theory of Colour (Zur Farbenlehre)* Div. 6, Historical Part. WA II.4; trans. Wells p.112.
10. from 'The Experiment as Mediator' *(Der Versuch als Vermittler,* 1823) para 28. HA 13.18; trans. Miller p.16.
11. MR 648 trans. Miller p.310.
12. MR 649 trans. Miller p.310.
13a. MR 646 trans. Miller p.310.
13b. MR 647 trans. Miller p.310.
14. MR 650 trans. Miller p.311 (adapted).
15. from *'Über Mathematik und deren Missbrauch'* (1826) GnS II.45; trans. Hjalmar Hegge, in Amrine p.201.

5. DELICATE EMPIRICISM
1. MR 509 trans. Miller p.307 2. MR 498 trans. Miller p.307.
3. from 'Problems' *(Probleme,* 1823). HA 13.37; trans. Miller p.44 4. MR
    512 trans. Miller p.307 (adapted).
5. MR 513 trans. Miller p.308.
6. MR 516 trans. Miller p.308.
7. MR 575 trans. Miller p.309.
8. from 'Significant Help' *(Bedeutende Fördernis durch ein einziges
    geistreiches Wort,* 1823). HA 13.40–41; trans. Miller p.41.
9. 'The Experiment as Mediator Between Subject and Object' *(Der Versuch
    als Vermittler von Objekt und Subjekt,* 1823). HA 13.10–20; trans.
    Miller pp.11–17.
10. from *'Einwirkung der neueren Philosophie.'* WA II.11.48–49; trans.
    Heinemann p.73.
11. Letter to Zelter 5/10/1828 trans. Coleridge p.334.
12. Letter to Sömmering 17/8/1795 trans. Weigand p.115.
13. from 'Towards a Theory of Weather' *(Versuch einer Witterungslehre,*
    1825). HA 13.312;1825) trans. Miller p.149.
14. GnS IV–2 p.358 trans. Heinemann p.68.
15. MR 435 trans. Miller p.306.
16. MR 504 trans. Miller p.307 (adapted).
17. MR 547 trans. Miller p.308.
18. MuR 1221 trans. Weigand p.123.
19. MuR 1222 trans. Weigand p.123.
20. MuR 529 trans. Stephenson 1983 p.114.
21. Letter to Zelter 5/10/1828 trans. Coleridge p.334.
22. 'Our Undertaking is Defended' *(Das Unternehmen wird Entschuldigt,*
    1817). HA 13.53–54; trans. Mueller p.21.

6. THE CONTEMPLATION OF NATURE SUGGESTS IDEAS
1. Letter to Steffens 29/5/1801 trans. Weigand p.95.
2. Preface to the *Theory of Colour (Zur Farbenlehre)* HA 13.317; trans.
    Miller p.159.
3. from 'Polarity' *(Polarität,* 1789). WA II.11.164; trans. Miller p.155.
4. MuR 199 trans. Stephenson 1983 p.99 5. Letter to Boisserée 25/2/1832
    trans. Heinemann p.75.
6. MR 488 trans. Miller p,.307.
7. MR 490 trans. Miller p.307.
8. Conversation with Reimar 19/3/1807 G.G. I.479 trans. Nisbet p.6.
9. MuR 558 trans. Stephenson 1983 p.45–46.
10. MuR 568 trans. Stephenson 1983 p.188.
11. WA I.2.216 trans. Nisbet p.20.
12. 'On the Spiral Tendency in Plants' *(Uber die Spiraltendenz der Vegetation,*

1831). WA II.7.37–68; trans. Vietor p.56.
13. from 'The Experiment as Mediator' (see Sources 5.9) trans. Miller p.15.
14. LA I.8.232; trans. Nisbet p.8.
15. 'Propitious Encounter' *(Glückiches Ereignis,* 1817) HA 10.538–42 ; trans. Mueller pp.215–19.
16. 'Indecision and Surrender' *(Bedenken und Ergebung,* 1820) HA 13.31–32; trans. Mueller pp.219–20.
17. 'Judgment through Intuitive Perception' *(Anschauende Urteilskraft,* 1820) HA 13.30–31; trans. Miller pp.31–32 (adapted).
18. from 'Significant Help' *(Bedeutende Fördernis,* 1823) HA 13.41–42; trans. Miller pp.39–40 (adapted).

7. THE ARCHETYPAL PHENOMENON
1. 'Experience and Science' *(Erfahrung und Wissenschaft,* 1718) HA 13.23–25; trans. Mueller pp.227–28 (adapted).
2. Letter to von Buttel 3/5/1827.
3. *Theory of Colour (Zur Farbenlehre)* II, section 175. HA 13.367; trans. Miller pp.194–95.
4. *Theory of Colour (Zur Farbenlehre)* II, section 177. HA 13.368; trans. Miller p.195.
5. Conversations with Eckermann 15/4/1827 trans. Weigand p.226.
6. MR 15 trans. Miller p.303.
7. MR 16 trans. Miller p.303.
8. MR 17 trans. Miller p.303.
9. Conversations with Eckermann 18/2/1829 trans. Weigand p.93 (adapted).
10. MR 577 trans. Sepper p.174.
11. Conversations with Müller 7/5/1830. CWBG 23.692; trans. Weigand p.76.
12. Conversations with Eckermann 13/2/1829 trans. Oxenford p.293–94 (adapted).

8. GOD IN NATURE, NATURE IN GOD
1. Annals (1811) JA 30.265; trans. Weigand p.74.
2. Letter to Jacobi 9/6/1785 trans. Weigand p.78.
3. MR 3 trans. Miller p.303.
4. *'Jacobis auserlesener Briefwechsel'* (1827) JA.38.125 trans. Vietor p.80.
5. Letter to Jacobi 5/5/1786 trans. Wells p.23–24.
6. *Dichtung und Warheit,* VI para. 10 trans. Blackie p.187.
7. Letter to Zelter 13/8/1831 trans. Coleridge p.458.
8. from *'Allgegenwärtig'* ('Her Omnipresence') D.L. 255; trans. Gray p.121.
9. from 'Indecision and Surrender' *(Bedenken und Ergebung,* 1820) HA 13.31; trans. Mueller p.219.
10. from *'Eins und Alles'* ('One and All') D.L. 275; trans. Luke p.275.
11. from *'Vermächtnis'* ('Legacy') D.L. 276; trans. Luke p.276.
12. Letter to the Duchess Louise, 23/12/1786; trans. Wannamaker.

9. SEEING AND SEEING

1. from 'Discovery of a Worthy Forerunner' *(Entdeckung eines trefflichen Vorarbeiters,* 1817) LA I.9.74; trans. Lehrs p.89.
2. from 'Significant Help' *(Bedeutende Fördernis)* HA 13.38; trans. Miller p.39.
3. *Theory of Colour (Zur Farbenlehre)* HA 14.41–42; trans. Sepper pp.194–95 (adapted).
4. *Theory of Colour, op.cit.* Preface, HA 13.315–16; trans. Miller p.158 (adapted).
5. from 'Ernst Stiedenroth: A Psychology in Clarification of Phenomena of the Soul' (1824) HA 13.42; trans. Miller p.45–46.
6. Conversations with Eckermann 27/1/1830 trans. Oxenford p.346.
7. WA IV.34.136–37; trans. Vasco p.88..
8. from introduction to *The Propyläea* (1798) HA 12.51; trans. Stephenson 1984 p.65.
9. from 'Preliminary Notes for a Physiology of Plants' *(Vorarbeiten zu einer Physiologie der Pflanzen),* 1790's, trans. Stephenson 1984 p.61.
10. from 'Review of Purkinje: Das Gehen in Subjectiver Hinsicht' (1819) GWBG XVI.902.
11. 'Der Kammerberg bei Eger' (1) (1809) HA 268; trans. Stephenson 1984 p.65–66 (adapted).
12. Letter to Jacobi (no date) trans. Heinemann p.79.
13. MR 364 trans. Wells p.24 (adapted).

10. THE HORIZON BOUNDLESS

1. 'A Friendly Greeting' *(Freundlicher Zuruf,* 1820), HA·13.34–35; trans. Miller p.37–38.
2. MuR 1080; trans. Stephenson 1983 p.92.
3. MR 514; trans. Miller p.308.
4. Conversations with Eckermann 26/2/1824; trans. Weigand p.124.
5. Conversations with Farthey 28/8/1827 GWBG 23.492; trans. Weigand p.136.
6. 'Epirrhema,' (1820) D.L. 273; trans. Roszak p.344.
7. Conversations with Riemar 2/8/1807; G.G. I.505.
8. MuR 546; trans. Stephenson 1983 p.117.
9. from 'A Study Based on Spinoza' *(Studie nach Spinoza,* c.1793) HA 13.7; trans. Miller p.8.
10. *ibid.*
11. from 'A Study Based on Spinoza,' *op.cit.* HA 13.8; p.9.
12. from 'Luke Howard to Goethe: a Bibliographical Sketch' (1822) HA 13.304; trans. Miller p.143.
13. from 'A Study Based on Spinoza,' *loc.cit.* p.9.
14. from 'Toward a Theory of Weather' *(Versuch einer Witterungeslehre,* 1825) HA 13.305; trans. Miller p.145.

15. from 'A Study based on Spinoza,' *loc.cit.* p.9.
16. Conversations with Eckermann 11/4/1827; trans. Weigand p.92–93.
17. from 'Karl Wilhelm Nose' WA II.9.195; trans. Norden p.239–40 (adapted).
18. MuR 1207.

Bibliography

The following short bibliography is intended to indicate just a handful of books which the reader interested in studying Goethe's approach to science further may find useful. There is a vast literature on this subject, and the reader wanting an extensive bibliography is referred to Fred Amrine's *Goethe and the Sciences: a Reappraisal* (1987) pp.389–437.

TEXTUAL SOURCES:
Miller, D. (ed.) *Goethe: Scientific Studies* (New York, Suhrkamp, 1988. New
      edition, Princeton University Press, 1995)
Mueller, B. (ed.) *Goethe's Botanical Writings* (Honolulu, University of Hawaii
      Press, 1952).

STUDIES:
Amrine, F. (ed.) *Goethe and the Sciences: a Reappraisal* (Dordrecht, D. Reidel
      Publishing Co., 1987).
Bortoft, H. *Goethe's Scientific Consciousness* (London, Institute for Cultural
      Research, 1986).
—, *The Wholeness of Nature: Goethe's Way of Science* (Edinburgh, Floris
      Books, 1996; New York, Lindisfarne 1996).
Lehrs, E. *Man or Matter* (London, Rudolf Steiner Press, 1985).
Magnus, R. *Goethe as a Scientist* (New York, Henry Schuman, 1949).
Nisbet, H.B. *Goethe and the Scientific Tradition* (London, Institute of Germanic
      Studies, 1952).
Sepper, D.L. *Goethe contra Newton* (Cambridge University Press 1988).
Steiner, R. *Goethean Science* (New York, Mercury Press, 1988).
Uberoi, J.P.S. *The Other Mind of Europe: Goethe as a Scientist* (Delhi, Oxford
      University Press, 1984).
Vietor, K. *Goethe the Thinker* (Cambridge, Mass.; Harvard University Press,
      1950).
Wells, G.A. *Goethe and the Development of Science: 1750–1900* (Alphen, 1978).

Index

The Wholeness of Nature

Goethe's Way of Science

Henri Bortoft

The approach of modern science is largely detached, intellectual and analytical. By contrast, Goethe's way of science pursued understanding through the experience of the 'authentic wholeness' of what was observed. Working with the intuitive mode of consciousness, Goethe aimed at an encounter with the whole phenomenon in its relationship with the observer. In his way of seeing, rather than dividing merely in order to categorize, we should investigate the parts of an object in order to reveal the true nature of the whole.

In this invaluable study, Henri Bortoft examines the phenomenological and cultural roots of Goethe's way of science. He argues that Goethe's insights, far from belonging to the past, represent the foundation for a future science. This new science of nature, involving other human faculties besides the analytical mind, can provide understanding and explanation in a way which our present scientific attitudes, and the culture they serve, desperately lack.

Henri Bortoft has taught physics and the philosophy of science for most of his career. His postgraduate research was on the problem of wholeness in the quantum theory under David Bohm and Basil Hiley at Birkbeck College, London. He now lectures and gives seminars on Goethean science as well as on the development of modern scientific consciousness.

Floris Books
Lindisfarne Press